BOTANICAL RESEARCH AND PRACTICES

MYCORRHIZAS

STRUCTURE, DEVELOPMENT AND FUNCTIONS

BOTANICAL RESEARCH AND PRACTICES

Additional books in this series can be found on Nova's website under the Series tab.

Additional e-books in this series can be found on Nova's website under the e-book tab.

BOTANICAL RESEARCH AND PRACTICES

MYCORRHIZAS

STRUCTURE, DEVELOPMENT AND FUNCTIONS

ELAINE WARWICK
EDITOR

nova publishers
New York

For permission to use material from this book please contact us:
Telephone 631-231-7269; Fax 631-231-8175
Web Site: http://www.novapublishers.com

Library of Congress Cataloging-in-Publication Data

Mycorrhizas : structure, development and functions / editor: Elaine Warwick.
pages cm. -- (Botanical research and practices)
ISBN: 978-1-63321-436-1 (hardcover)
1. Mycorrhizal fungi. 2. Mycorrhizas. I. Warwick, Elaine. II. Series: Botanical research and practices.
QK604.2.M92M9653 2014
631.4'6--dc23

2014023752

Published by Nova Science Publishers, Inc. † New York

CONTENTS

PREFACE

Mycorrhizae represent a symbiotic relationship between a fungus and plant roots. The fungus colonizes the roots of the host plant, either intracellularly in arbuscular mycorrhizal fungi, or extracellularly in ectomycorrhizal fungi. Host plants enhance the defense of mycorrhizal fungi against fungivores. Mycorrhizae mediate host plant-insect interactions. Arbuscular mycorrhiza play a crucial role in augmenting plant resistance to stressful conditions like diseases, drought, heavy metal toxicity, high temperatures, salinity, and waterlogging. They are also important in enhancing decomposition of toxic organic substances, in remediating polluted environments, and in keeping a stable ecosystem and sustainable productivity. This book discusses the structure, development and function of mycorrhizas.

Chapter 1 – The establishment and the maintenance of a mutualistic arbuscular mycorrhizal (AM) interaction are based on intense communications between the partners and within and between plant cells and organs. Legumes, which can in addition to the AM symbiosis undergo a symbiosis with nitrogen-fixing rhizobacteria, serve as model plants to study the signaling in both root endosymbioses. This chapter will summarize the current knowledge on early signaling and regulation of arbuscular mycorrhization as follows. The first part provides an overview on the signaling leading to successful infection of the root – including common elements required for both legume endosymbioses. This part comprises (i) signals involved in attraction and recognition of the partners, (ii) the activation of a signaling cascade in root hair cells of legumes by AM fungal and rhizobacterial signals, and (iii) the mechanisms leading to successful penetration of the root surface and to intracellular accommodation of the AM fungal partner. The second part focuses on the carbon supply of the fungus and on the role of apoplastic invertases and sugar-induced signaling in the AM interaction. The regulation of the AM symbiosis by a long-distance and systemic feedback mechanism, called autoregulation of mycorrhization, is topic of the last part of this chapter.

Chapter 2 – Most plant species associate with mycorrhizas to perform better under stressful conditions and the increasing recognition of the mycorrhizal interactions, especially arbuscular mycorrhizal fungi (AMF) may have important functions in terrestrial ecosystems. AMF structures such as total root colonized and percent root length colonized by arbuscules can be used to estimate of AMF abundance within a plant or ecosystem. Also important is to evaluate the extraradical hyphae of AMF that extend into the surrounding soil much further than do roots. Due to their strategic position at the soil–root interface, the study of AMF in different undisturbed or degraded ecosystems can contribute to better understand the effect of global change on plant-soil associated communities. Many plant species in tropical regions where semiarid environments are also found associate with AMF and this symbioses has to be monitored under global change. This chapter explores the current available information relevant to benefits from AMF symbioses in natural and disturbed sites. Possible functions of AMF structures and research on Glomalin and AMF interaction with biochar are also discussed. Better knowledge of the wide variation in AMF parameters is important for understanding the ecology of different land ecosystems and for management purposes.

Chapter 3 – Mycorrhizae represent a symbiotic relationship between a fungus and plant roots. The fungus colonizes the roots of the host plant, either intracellularly in arbuscular mycorrhizal fungi, or extracellularly in ectomycorrhizal fungi. Host plants enhance the defense of mycorrhizal fungi against fungivores. Mycorrhizae mediate host plant-insect interactions.

Arbuscular mycorrhizal exudates are important to the formation of lateral roots, and in the absorption and accumulation of elements including manganese, phosphorus, potassium, and zinc. Mycorrhizae alter nitrogen acquisition and improve plant growth. Organic nitrogen is more accessible to etomycorrhizal and ericoid mycorrhizal fungi than arbuscular mycorrhizal fungi because of the presence of enzymes capable of degrading nitrogen-containing molecules in the former type of fungi. Ectomycorrhizal fungi are involved in decomposing and mobilizing nitrogen from organic matter-protein complexes in forest soils. Ectomycorrhizae induce changes in the surface structure and composition of soil colloids.

Mycorrhizae are involved in phytoremediation of soil contaminated with heavy metals. They also enhance plant tolerance to adverse conditions, decrease aluminium phytotoxicity in acidic soils, and improve plant antioxidant systems and tolerance to salinity.

Chapter 4 – In soil N is present in inorganic forms as ammonium (NH_4^+) and nitrate (NO_3^-), and as organic compounds such as amino acids, peptides and proteins. However, availability of N appears to be a limiting factor for forest tree growth in most temperate and boreal ecosystems where a significant pool of N is present in complex organic compounds.

Ectomycorrhizal fungi play a major role in soil N cycling by their utilization of both inorganic and organic forms. Nutrients, including N, are mobilized from soil by fungal mycelia and transferred to the host plant.

On the other hand, the host feeds the fungal symbiont with carbon compounds derived from photosynthesis. Despite the widely accepted importance of ectomycorrhizas on forest tree nutrition and N cycling, the control of basic N metabolism in ectomycorrhizal fungi is still poorly studied. Therefore, resolving how ectomycorrhizal fungi regulate their N adquisition will contribute to a better understanding of forest tree nutrition. *Laccaria bicolor* is an ectomycorrhizal basidiomycete which forms symbiosis with several boreal and temperate forest trees such as birch, pine and poplar.

Ectomycorrhizal fungi such as *L. bicolor* compete with saprotrophs and plants for the limited pool of nutrients. Ectomycorrhizal fungi are well known to utilize complex organic N sources especially abundant in acid boreal and temperate forest soils but mineral N is efficiently utilized as well. Ectomycorrhizal fungi appear to be well adapted to low mineral N concentrations and efficiently assimilate inorganic N as NH_4^+, NO_3^-, the first generally being the preferred N source. While NH_4^+ is the dominant mineral N form in undisturbed boreal and temperate forest soils the true availability of NO_3^- for ectomycorrhizal fungi in their habitat is not well established. The majority of studied ectomycorrhizal fungi shows a conserved capacity to grow on NO_3^- as the sole N source and some of them have even been reported to prefer NO_3^- over NH_4^+. This suggests that NO_3^- may play a more significant role as N source in forest ecosystems as traditionally believed. *Laccaria*, like many other ectomycorrhizal fungi, clearly prefers NH_4^+ but it also grows well on NO_3^-. Its genome encodes for one copy of each NO_3^- assimilation genes: the transporter *Lbnrt,* the NO_3^--reductase *Lbnr* and the NO_2^- -reductase *Lbnir*. The genes are organized in the so-called fungal high-affinity N-assimilation cluster (fHANT-AC). The authors found massive effects on the formation of ECM root tips of poplar when N-assimilation in *L. bicolor* was impaired by silencing of a selected NO_3^--metabolism gene, the nitrate reductase encoding gene *Lbnr*. Alteration of the symbiotic capacity was specific to the RNAi-induced reduction in N assimilation of the fungal partner. Silencing of *Lbnr* strongly impaired mycorrhiza formation on NO_3^- but not on NH_4^+. The participation of NO_3^- -utilization genes in the symbiotic interaction of *Laccaria* is hence clear.

The authors have extended the nitrate utilization RNAi-based gene silencing studies of the ECM basidiomycete *Laccaria bicolor* by altering the expression of *LbNrt*, the sole nitrate transporter encoding gene of the fungus. Here they report the first nutrient transporter mutants for mycorrhizal fungi. Silencing of *LbNrt* results in fungal strains with minimal detectable *LbNrt* transcript levels, significantly reduced growth capacity on nitrate and altered symbiotic interaction with poplar. Transporter silencing also produces co-downregulation of the whole fHANT-AC. Moreover, this effect on the nitrate utilization pathway seems to be independent of extracellular nitrate or the nitrogen status of the fungus. The authors' results show a novel and key nitrate uptake-independent regulatory role for a eukaryotic nitrate transporter. The possible cellular mechanisms underneath this regulation are discussed in the light of the current knowledge on NRT2-type nitrate transporters.

Chapter 5 – The capacity of roots to sense soil physicochemical parameters plays an essential role in maintaining plant nutritional and developmental functions under abiotic stress. Drought greatly affects the growth and development of plants and can cause extensive losses to agricultural productivity. The water regime affects a wide variety of physiological and biochemical processes in plants including an increased production of reactive oxygen species (ROS) capable of causing oxidative damage to proteins, DNA and lipids. Moreover, this ROS production can induce cellular, anatomical, and morphological changes that improve drought tolerance. Arbuscular mycorrhizal fungi (AMF) colonize a wide range of plant species though the ability of different AMF to promote host growth and/or water deficit resistance is variable. AMF contributes to drought tolerance through a combination of physical, nutritional and physiological effects. The response to oxidative stress depends on many factors, including the organism tissue and the degree of stress. In most cases mycorrhizal plants increase antioxidant defenses, such as the ROS-scavenging enzymes superoxide dismutase (SOD), catalase (CAT), and ascorbate peroxidase (APX). AMF can maintain the levels of malondialdehyde (MDA), a compound indicative of lipid damage, by rapidly increasing antioxidant defenses and preventing lipid damage. In other cases an increase of MDA is observed in mycorrhizal plants due to an active response to oxidative stress. A critical step of AMF inoculation technology is the appropriate selection of fungal isolates to be used as inoculants. Nursery cultivation is the most convenient way to improve the success of semi-woody cuttings transplantation (e.g. olive, pomegranate, citrus) at the crop area. Plants suffer at least two transplant moments in nursery conditions prior to outside cultivation. AMF can reduce transplant stress by changing the morphology of the root system favouring the establishment of plants. Thus AMF can greatly contribute to crop productivity and environmental sustainability. Artificial inoculation of plant cuttings with AMF at an early stage of cutting development has been adopted by an increasing number of nursery managers as a method for promoting growth, production and precocity. Moreover, it is an essential component for most plants and it can be used as a biofertilizer resource. In this chapter the authors discuss the effect of an early AMF inoculation in alleviating oxidative stress in semi-woody cutting propagation management under nursery conditions. They also discuss the benefits of AMF to solve the problematic issue of water deficit and transplant stresses generated during plant growth.

Chapter 6 – This article reviews the state of the art with respect to the environmental effects of wastewater irrigated agriculture on arbuscular mycorrhizal fungi in arid and semi-arid regions on a field scale. Petroleum, tannery and pharmaceutical compounds emerge as pollutants when wastewater is used for irrigation without any previous treatment, situation which is now the object of new studies. In the last few decades, the amounts of detergents, fats, oils, heavy metals and synthesized organic compounds generated by industrial activities have increased, and some 10,000 new organic compounds are added each year. and released into the environment for direct or indirect application over a long period of time. In both developed and developing countries, the most prevalent practice is the application of municipal wastewater (both treated and untreated) to land. Impact from wastewater on agricultural soil, is mainly due to the presence of high nutrient contents (Nitrogen and Phosphorus), high total dissolved solids and compounds pollutants, which are added to the soil over time. Arbuscular mycorrhizal plants have two pathways that can contribute to nutrient uptake from soil; direct uptake from the rhizosphere by root epidermal cells and the vesicular arbuscular mycorrhyzae uptake pathway via the fungi. The establishment of the vesicular arbuscular mycorrhizae symbiosis can be disturbed by environmental conditions, including phosphorus (P) availability which can inhibit the symbiotic interaction. Interestingly, recent studies have demonstrated a cross-talk between phosphorus and nitrogen availabilities to control arbuscular mycorrhizal associations, thus indicating a high level of integration of mycorrhizal responses to mineral nutrition. Some authors have reported a direct effect of P on arbuscular mycorrhizal spore germination and hyphal growth. Evidence has also been gathered for downregulation by P of well established vesicular arbuscular mycorrhizae interactions, via a reduced production of root compounds. Arbuscular mycorrhizal symbiosis could be arrested almost completely by a high P supply at a very early stage, prior to the attachment of the fungus on the root. The synthesis and exudation of strigolactones are negatively affected by a high availability of P. Under conditions of P sufficiency, a strong reduction of strigolactone release would prevent the stimulation of AM fungi and hence the establishment of the interaction. The impact of arbuscular mycorrhizas, is starting to draw attention in the context of wastewater irrigation and phytoremediation of organics, and recent results have shown that these symbioses may be crucial for both plant establishment on contaminated sites.

Chapter 7 – Fusarium root rot tolerance and changes in antioxidative ability in arbuscular mycorrhizal fungi (AMF) [*Glomus intraradices* (Gi) and *Gigaspora margarita* (GM)] and non-pathogenic Fusarium (NPF) [*Fusarium oxysporum* f. sp. asparagi (Foa)]-inoculated asparagus plants (*Asparagus officinalis* L., cv. 'Welcome') were investigated. Dry weight of shoots and roots increased in AMF and NPF-inoculated plants. Though, dry weight increased in NPF-inoculated plants compared than control, which is comparatively lower than AMF-inoculated plants. On the other hand, SOD activity, APX, DPPH radical scavenging activity, content of polyphenol and ascorbic acid increased under inoculation of pathogen stress condition; these effects differed between shoots and roots. As for disease tolerance, the incidence and severity of symptom was less in both AMF and NPF-inoculated asparagus plants than non-inoculated plants, though the effect differed between Gi and GM inoculated asparagus plants with or without presence of NPF. These results suggest that tolerance to Fusarium root rot was increased by AMF colonization with or without NPF in asparagus plants. As well as, antioxidative ability increased in pathogen-stressed conditions, so that the Fusarium root rot tolerance might be associated with the increase in such antioxidative factors.

In: Mycorrhizas
Editor: Elaine Warwick

ISBN: 978-1-63321-436-1

Chapter 1

MOLECULAR SIGNALING IN THE ARBUSCULAR MYCORRHIZAL SYMBIOSIS

Sara Schaarschmidt[1*] ***and Peter M. Gresshoff***[2†]
[1]Humboldt-Universität zu Berlin, Germany
[2]The University of Queensland, St Lucia, Brisbane, Australia

ABSTRACT

The establishment and the maintenance of a mutualistic arbuscular mycorrhizal (AM) interaction are based on intense communications between the partners and within and between plant cells and organs. Legumes, which can in addition to the AM symbiosis undergo a symbiosis with nitrogen-fixing rhizobacteria, serve as model plants to study the signaling in both root endosymbioses. This chapter will summarize the current knowledge on early signaling and regulation of arbuscular mycorrhization as follows. The first part provides an overview on the signaling leading to successful infection of the root – including common elements required for both legume endosymbioses. This part comprises (i) signals involved in attraction and recognition of the partners, (ii) the activation of a signaling cascade in root hair cells of legumes by AM fungal and rhizobacterial signals, and (iii) the mechanisms leading to successful penetration of the root surface and to

[*] E-mail: sara.schaarschmidt@hu-berlin.de.
[†] E-mail: p.gresshoff@uq.edu.au.

intracellular accommodation of the AM fungal partner. The second part focuses on the carbon supply of the fungus and on the role of apoplastic invertases and sugar-induced signaling in the AM interaction. The regulation of the AM symbiosis by a long-distance and systemic feedback mechanism, called autoregulation of mycorrhization, is topic of the last part of this chapter.

1. INTRODUCTION

Terrestrial plants evolved different strategies to improve their supply with mineral nutrients – one comprises the formation of mycorrhizas, mutualistic interactions with soil-born fungi. The arbuscular mycorrhizal (AM) interaction represents the most wide-spread and oldest type of mycorrhizas, and the oldest plant–microbe symbiosis at all. The origin of the AM fungi, which form the phylum Glomeromycota (Schüßler et al. 2001, Schüßler & Walker 2010), is dated back to more than 600 million years ago (MYA) (Schüßler & Walker 2011). The fungi are hypothesized to facilitate the colonization of land by plants (for review see Brundrett 2002, Strullu-Derrien & Strullu 2007, Bonfante & Genre 2008, Corradi & Bonfante 2012). Fossil records proof the presence of Glomalean fungi in the Ordovician (around 400 MYA) (Redecker et al. 2000) and of AM-like endosymbiotic structures in early land plants of the Devonian (around 460 MYA) (Remy et al. 1994). AM fungi are believed to colonize plants even before those evolved roots supplying water and mineral nutrients to the early terrestrial plants and improving their fitness as recently demonstrated for liverworts (Humphreys et al. 2010). Today, most terrestrial plants world-wide, including non-vascular plants, herbaceous and wooden plants, and many important crops, are mycotrophs still relying on the benefits provided by AM fungi. The main characteristic of the symbiosis is the nutrient exchange between the partners. Whereas the obligate biotrophic AM fungi receive carbon from plants, the plant partner profits by an improved supply with minerals, mainly phosphate, and water. In addition, mycorrhizal plants often exhibit an improved abiotic and biotic stress tolerance. By the overall increased plant fitness and competitiveness, the AM symbiosis is a major factor determining ecosystem biodiversity and productivity (van der Heijden et al. 1998, Wagg et al. 2014) and is essential for sustainable agriculture (Gianinazzi et al. 2010).

The arbuscular mycorrhiza is an endomycorrhiza with fungal structures formed within the root cells. The exchange of nutrients occurs via the intraradical fungal structures. Different morphological types of the AM symbiosis can be distinguished with two main types named *Paris* and *Arum* type (for overview see Smith & Smith 1997, Dickson et al. 2007). Particularly in ferns, gymnosperms and many wild angiosperms, a *Paris* type is formed. Here the AM fungi grow in an intracellular way along the root with coiled intracellular hyphae that are the main organ for the nutrient exchange in the *Paris* type (Figure 1a). Hyphal coils can feature branching forming arbuscules (arbusculate coils). In common crops, AM fungi mostly form an *Arum* type (Figure 1b), in which the AM fungi spread along the root with hyphae formed between the root cortex cells along root air channels. Emanating from these intercellular hyphae, the fungi grow into the cortex cells and form by multitudinous branching events the typical arbuscules (Figure 1). In contrast to hyphal coils, arbuscules are always terminal structures and are thus characteristic for the *Arum* type. The highly-branched (tree-like) arbuscules have a strongly increased surface and are the fungal organ for mineral nutrient exchange in the *Arum* type. They have a limited life-span and finally collapse and undergo degradation (Cox, Alexander et al. 1988) allowing new arbuscule formation in the same cell. In addition to the *Arum* and *Paris* type, intermediate morphological forms can occur (for structural diversity). Although AM morphology appears to be mainly determined by the host, the AM fungi also exhibit a certain impact (Smith & Smith 1997, Dickson 2004). In some plant species, both morphological types can be formed as once when the plant is colonized by different fungal taxa (Kubota et al. 2005). However, the underlying mechanisms resulting in a *Paris*, *Arum* or intermediate type are so far unknown. Although large continuous airspaces in the root cortex seem to be essential for the formation of intercellular hyphae, air channels are no exclusion criterion for a *Paris* type (Dickson et al. 2007). So far, most AM research is done on the *Arum* type which is also the predominating morphological type in legumes.

Establishment and maintenance of the AM endosymbiosis goes along with severe transcriptional, cellular, and metabolic adaptations of the host plant to enable successful attraction, recognition, intracellular accommodation, and control of the microsymbiont. In contrast to the obligate biotrophy of the AM fungi, plants are not necessarily dependent on the AM interaction. The AM symbiosis is particularly determined by the phosphate availability for the plant but also by other unfavorable conditions in which the plant can benefit from the fungus and which are usually or quite frequently present in ecosystems, such as limitation in other mineral nutrients and/or water, elevated soil salinity, heavy metal contamination, and biotic stress conditions including competition. Here, an overview is given on signals and mechanisms underlying the processes of successful infection and colonization of a host root by AM fungi as well as regulatory mechanisms that optimize carbon efficiency for the host and influence the mycorrhization pattern of the plant like the autoregulation of mycorrhization.

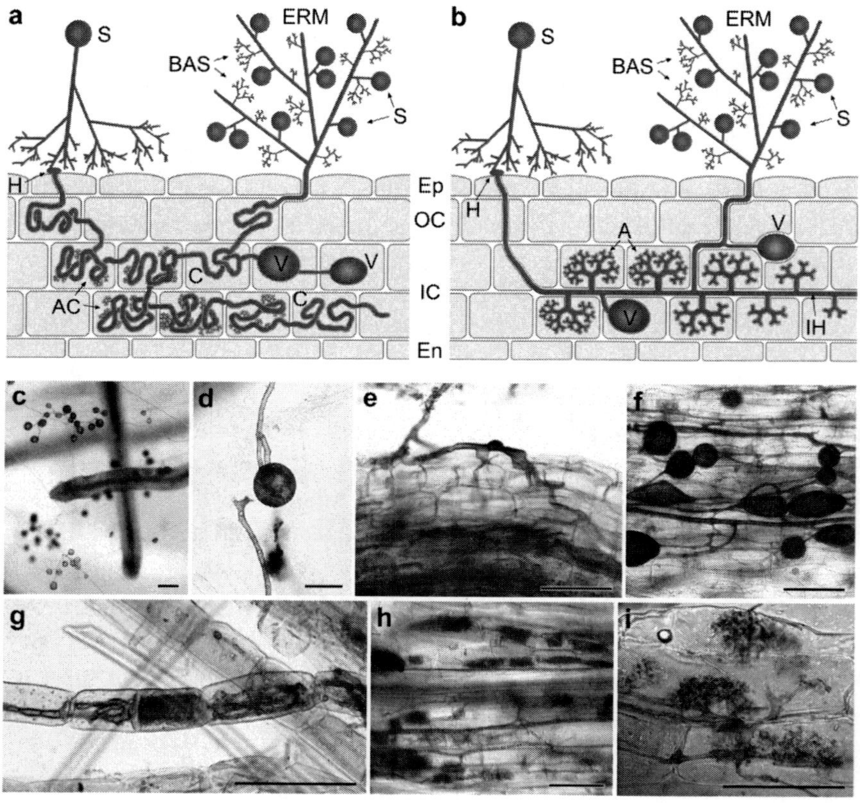

Figure 1. Morphological structures of the AM symbiosis. a+b: Schematic presentation of the *Paris* (a) and the *Arum* (b) type. c: Extraradical mycelium with spores. d: Single Glomalean spore. e: Hyphopodium and entry point. f: Vesicles. g: Intraradical coils and arbusculate coils. h: Intercellular hyphae and arbuscules. i: Arbuscules. Fungal structures in d-i are stained according to Vierheilig and co-workers). Bars represent 250 µm (c) and 50 µm (d-i). S: spore; H: hyphopodium; C: coil; AC: arbusculate coil; A: arbuscule; IH: intercellular hypha; V: vesicle; ERM: extraradical mycelium; BAS: branched adsorbing structures.

Plant genes required for establishment and regulation of an AM symbiosis were often identified by plant mutant analyses. Such analyses can in addition reveal common symbiosis genes involved in the AM symbiosis as well as in the interaction of plants with nitrogen-fixing bacteria, which developed later during evolution and comprises the legume–rhizobia symbiosis (LRS) also called nodulation.

Root-specific processes can be also well investigated in composite plants consisting of wild-type shoots and transgenic roots achieved by *Agrobacterium rhizogenes*-mediated root transformation or, if no shoot-derived signal(s) are required, in root organ cultures. Root cultures can be established with either non-transformed roots or with roots transformed with (wild-type) *A. rhizogenes* resulting in so called hairy roots, which are characterized by a higher growth potential. Cultivation on growth medium with lower mineral contents allows successful mycorrhization of those root cultures.

In contrast, for analyses of the LRS, such root cultures are less suitable. Here, the bacteria are well able to utilize the sugars provided in the medium without forming an LRS; moreover, nodulation is inhibited by nitrogenous compounds such as nitrate. However, for the symbiosis with the obligate biotrophic AM fungi, root organ cultures are a powerful tool to analyze this symbiosis under *in vitro* conditions including investigations of fungal-specific processes and metabolites during the symbiotic phase by obtaining pure, metabolic-active AM fungal material in form of extraradical mycelium (for overview see Fortin et al. 2002).

Most research on AM molecular signaling has been done with few fungal species formerly classified as *Glomus* species (Schüßler & Walker 2010), including the AM model fungus *Rhizophagus irregularis* DAOM 197198 formerly known as *Glomus intraradices* DAOM 197198 (Stockinger et al. 2009, Schüßler & Walker 2010), and few *Gigaspora* species. So far, functional gene analyses in AM fungi are less applicable, but a first progress in introducing foreign DNA could be achieved via particle bombardment (Harrier & Millam 2001) and recently, gene suppression was shown to be possible by host-induced gene silencing (Helber et al. 2011). Moreover, recent sequencing of the transcriptome and genome of the model fungus *R. irregularis* DAOM 197198 will strongly improve our understanding of the AM symbiosis (Tisserant et al. 2012, 2013).

2. Establishment of an AM Symbiosis

The genetic basis for an AM establishment probably evolved with or even before the colonization of land by plants. Most of our today's land plants still hold the ability to interact with AM fungi to form a mutualistic interaction indicating the importance of this symbiosis for plant competitiveness and ecosystem functioning. However, some plant species lost this ability including members of the Brassicaceae. The initialization of an AM symbiosis is based on an intense signaling exchange between both partners, which is in parts also recruited for the LRS.

2.1. Attraction of AM Fungi and Perception of Specific Fungal Signals

To form a mutualistic interaction, the involved partners first of all have to find and recognize each other. Particularly under phosphate starvation, plants exude different signals into the rhizosphere to attract AM fungi (for overview see Figure 2). In the absence of a host, AM fungal spores are able to germinate, but show only limited hyphal development. If no plant-derived signals are perceived by the fungus, it stops asymbiotic growth and reallocates resources allowing multiple germinations (Koske 1981, Logi et al. 1998). Numerous studies showed the promoting effect of root exudates on spore germination and their requirement for the switch from asymbiotic to presymbiotic AM fungal growth by stimulating hyphal development and branching prior to physical contact of the partners (e.g. Graham 1982, Elias & Safir 1987, Poulin et al. 1993, Nagahashi & Douds 2007, Sun). Presymbiotic fungal development is preceded and accompanied by rapid induction of fungal gene expression and stimulated respiratory and mitotic activity of the fungus caused by root exuded compounds (Tamasloukht et al. 2003, Besserer et al. 2006, 2008). CO_2 emitted by the root as a volatile was demonstrated as critical factor for the stimulation of AM fungal growth by root exudates, probably by serving as essential carbon source (Bécard & Piché 1989).

2.1.1. Flavonoids

One group of root exuded compounds that can attract AM fungi are the flavonoids – secondary plant metabolites derived from the phenylpropanoid pathway (for overview see Cesco et al. 2012). Root flavonoids are well researched for leguminous plants since they promote root nodulation by stimulating the production of specific rhizobacterial signals the so-called Nod factors, but root flavonoids are ubiquitously distributed in the plant kingdom. For AM fungi it was shown that flavonoids can stimulate spore germination, hyphal growth and branching, and root infection and colonization (for overview see Steinkellner et al. 2007, Abdel-Lateif et al. 2012, Hassan & Mathesius 2012).

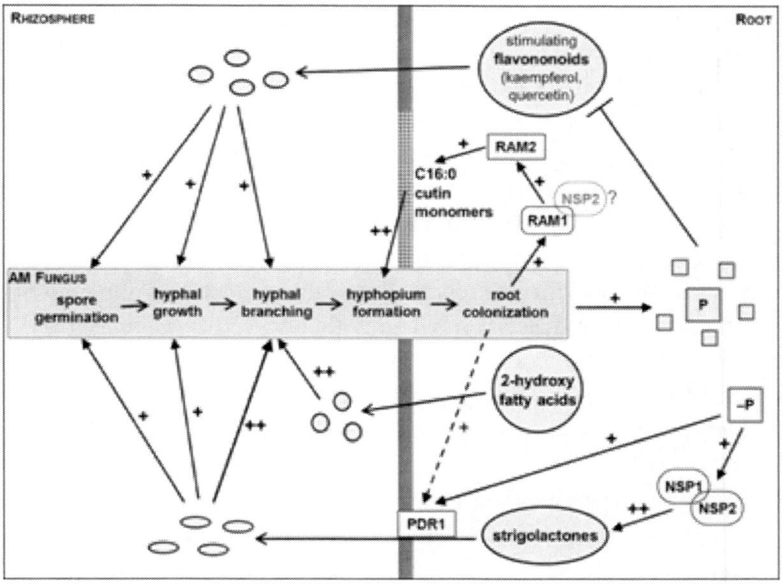

Figure 2. Signals involved in attraction of AM fungi. Different compounds exuded by roots stimulate germination of AM fungal spores, hyphal growth and branching. These factors include specific flavonoid compounds that in general induce germination and presymbiotic fungal development, 2-hydroxy fatty acids that stimulate elongated lateral branches, and strigolactones that activate fungal energy metabolism and that are particularly important for hyphal branching, also of higher order. Strigolactone biosynthesis upon phosphate (P) starvation is regulated by MtNSP1 and MtNSP2, GRAS domain transcription factors that are also essential for the Nod factor response (Liu et al. 2011). Exudation of strigolactones is controlled by an ABC transporter, PhPDR1 (Kretzschmar et al. 2012). Attraction of AM fungi by root exudates is regulated by the nutritional status of the plant. Additional factors, including volatiles, are involved in fungal attraction (not depicted). For hyphopodium formation, specific cutin monomers of the epidermal cell wall are essential signals, whose production is regulated by the GRAS domain transcription factor MtRAM1 (that can also interact with MtNSP1) and the glycerol-3-phosphate acyl transferase MtRAM2 (Gobbato et al. 2012, Wang et al. 2012). +: positive regulation; ++: essential factor for induction.

However, application experiments revealed flavonoid-specific and concentration-dependent effects on spore germination and/or presymbiotic growth of AM fungi and demonstrated that flavonoids can act as stimulators but also as inhibitors (for example see Siqueira et al. 1991, Tsai & Phillips 1991, Bécard et al. 1992, Morandi et al. 1992). A stimulating effect was particularly found for flavonols like kaempferol and quercitin, which could be detected in root exudates of legumes and non-legumes like carrot (*Daucus carota*) (Tsai & Phillips 1991, Bécard et al. 1992, Poulin et al. 1993, Scervino et al. 2005b). In contrast, some flavonoid compounds, e.g. the flavones apigenin, chrysin, and luteolin, were demonstrated to inhibit germination, presymbiotic growth, and/or entry points of AM fungi (Bécard et al. 1992, Scervino et al. 2007). Also pyranoisoflavones isolated from root exudates of the non-host plant white lupin (*Lupinus albus*), a member of the usually mycotrophic Fabaceae family, were identified as inhibitors of AM fungal germ tube growth (Akiyama et al. 2010b). With some isoflavones contradictory results were achieved. Whereas few studies found reduced germination of two *Glomus* species and reduced hyphal growth of *Gigaspora margarita* after application of formononetin and biochanin A, respectively (Tsai & Phillips 1991, Bécard et al. 1992), other studies showed improved AM fungal hyphal growth and root colonization by each of the two isoflavones for other *Glomus* species (Nair et al. 1991, Siqueira et al. 1991). This might be due to differences in the experimental setup including the flavonoid concentration but also to genus/species-specific differences that were found in the responsiveness of AM fungi towards individual flavonoids (Tsai & Phillips 1991, Scervino et al. 2005a, 2007).

In the rhizosphere, the particular potpourri of exuded flavonoids rather than individual flavonoid compounds would determine the effect on AM fungi. Interestingly, plants are able to quantitatively and qualitatively modulate their flavonoid exudation according to their nutritional and mycorrhizal status. For bean (*Phaseolus vulgaris*), root exudation of phenolic compounds including flavonoids was found to be strongly induced not only upon nitrogen starvation but also upon phosphate starvation (Juszczuk et al. 2004). In melon (*Cucumis melo*), a flavonoid that stimulates AM fungal colonization was found to accumulate in roots upon phosphate deficiency, whereas elevated phosphate fertilization and AM fungal inoculation led to a strongly reduced accumulation (Akiyama et al. 2002). However, although several flavonoids can stimulate AM fungal spore germination, presymbiotic fungal growth, and root infection and colonization, they are not essential for an AM interaction as demonstrated with flavonoid-deficient plants (Bécard et al. 1995). Thus, additional plant-derived signals must exist.

2.1.2. Strigolactones

In the year 2000, non-flavonoid compound(s) of root exudates of mycotrophic plants were identified to induce hyphal branching, nuclear division, and hyphal length of germinating AM fungi, whereas exudates of non-host plants had no effect (Buee et al. 2000, Nagahashi & Douds 2000). Five years later, the strigolactone 5-deoxy-strigol was purified as a branching factor from exudates of phosphate starved *Lotus japonicus*, leading to hyphal branching in the AM fungus *G. margarita* (Akiyama et al. 2005). Strigolactones are a new class of carotenoid-derived phytohormones that were first discovered in cotton root exudates as germination stimulants for the parasitic weed *Striga lutea* (Cook et al. 1966, 1972). During the last decade, their regulatory role in root and shoot development and their function for the interaction with AM fungi have been intensively studied (for review see e.g. Brewer et al. 2013, Rasmussen et al. 2013, Ruyter-Spira et al. 2013, Zwanenburg & Pospíšil 2013). In roots, which generally exhibit high levels of strigolactones, these compounds are involved in regulation of root architecture, probably by regulating lateral auxin reflux. They can inhibit adventitious root formation, stimulate primary root growth and root hair elongation (the latter upon involvement of ethylene), and affect lateral root initiation. Lateral root formation can be either reduced (under sufficient P availability) or improved (under P deprivation) by strigolactones. Strigolactones were also found to promote LRS formation as they somehow stimulate root nodule number in pea (Foo & Davies 2011, Foo et al. 2013b) and presumably activate swarming mobility of rhizobacteria (Tambalo et al. 2014).

In germinating *Gigaspora* species, not only the strigolactone 5-deoxy-strigol, but a whole range of natural strigolactone compounds, including sorgolactone and strigol found in root exudates of the monocot *Sorghum bicolor,* and several synthetic analogues, including GR24 and G7, induce AM hyphal branching at very low concentrations (Akiyama et al. 2005, 2010a, Besserer et al. 2006). Moreover, the branching response was also detected for phylogenetically distant AM fungi (Besserer et al. 2006). These findings reveal the common role of this phytohormone class as plant-derived signals in the widespread AM symbiosis. Strigolactones boost the presymbiotic development by stimulating the AM fungal metabolic activity including mitochondrial biogenesis and respiration (Besserer et al. 2006). Strigolactones rapidly induce the NADH content of the hyphae which can be used by the stimulated mitochondrial NADH dehydrogenase activity to mobilize the fungal lipids as carbon and energy source required for the stimulated cell proliferation (Besserer et al. 2008). The induced branching pattern can differ between individual strigolactones and comprises also highly branched short hyphae of higher order (Akiyama et al. 2010a).

Strigolactone exudation is well controlled by the plant. Recently, the first strigolactone transporter was identified (Kretzschmar et al. 2012). An ATP-binding cassette (ABC) transporter from petunia (*Petunia hybrida*), PhPDR1, was shown to control as cellular exporter the efflux of strigolactones from the plant root. If this transport is abolished, the AM fungal branching and the root colonization are reduced. The transporter is transcriptionally induced by phosphate starvation, application of G24 or auxin, and interestingly also upon AM interaction (Kretzschmar et al. 2012); however, another study showed reduced strigolactone activity of root exudates due to AM fungal root colonization (Lendzemo et al. 2009). Induction of the ABC transporter at phosphate deprivation is in line with promoted biosynthesis and elevated exudation of strigolactones upon P (and N) starvation in legumes and non-leguminous plants (Yoneyama et al. 2007a, 2007b, López-Ráez et al. 2008, Foo et al. 2013b). The biosynthesis of strigolactones is probably controlled by transcriptional regulators and recently, two GRAS domain transcription factors were found to be indispensable for strigolactone production under nutrient-deficiency conditions (Liu et al. 2011). The *Medicago truncatula* transcription factors Nodulation Signaling Pathway (NSP)1 and NSP2, that were previously identified to activate Nod factor-induced gene expression (Hirsch et al. 2009), and orthologs from rice (*Oryza sativa*) regulate expression of carotenoid biosynthesis genes and strigolactone accumulation in roots under P (and N) starvation (Liu et al. 2011). A negative regulation of strigolactone biosynthesis by high P supply occurs systemically in the root as shown for pea (*Pisum sativum*) (Balzergue et al. 2011). However, since exogenous application of strigolactones was not sufficient to restore mycorrhizal colonization under high-P conditions, additional factors other than strigolactone must be involved in mediating the effect of P availability on the AM symbiosis (Balzergue et al. 2011). This is consistent with analyses on pea mutant plants. These analyses revealed that neither strigolactone biosynthesis nor a strigolactone response is required for the enhanced mycorrhization under low-P conditions (Foo et al. 2013b).

Next to mycotrophic plants, also AM non-host plants exude strigolactones as shown for *L. albus*; here, however, exudation is not correlated to the nutritional status of the plant (Yoneyama et al. 2008). Astonishing, the AM fungal branching response to the lupin-derived strigolactones was completely inhibited by pyranoisoflavones exuded by *L. albus* (Akiyama et al. 2010b).

2.1.3. Additional Branching Factors

For the complex branching pattern of AM fungi in response to root exudates, a mixture of compounds is suggested as inducer instead of a single factor. Different alcohol-soluble fractions of *D. carota* root exudates were found to induce different branching phenotypes – also in dependence on the AM fungal species (Nagahashi & Douds 2007). In addition to flavonoids and strigolactones, 2-hydroxy fatty acids of specific lengths can act as branching signals (Nagahashi & Douds 2011). 2-Hydroxytetradecanoic acid and 2-hydroxydodecanoic acid from *D. carota* root exudates strongly increased elongated lateral branches of the primary germ tube and of major secondary hyphae. A methanol-soluble factor (not identical to GR24) that stimulates presymbiotic AM fungal growth, hyphal branching and AM establishment was also found in tomato (*Solanum lycopersicum*) root exudates (Sun) but its chemical structure is so far unknown. Mutant plants missing this factor emit soluble factor(s) inhibiting spore germination and hyphal tip growth and are defective in premycorrhizal infection (*pmi* mutant) (Gadkar et al. 2003). However, despite the inhibiting factor(s) of the mutant, application of the stimulating factor from wild-type tomato plants to the *pmi* mutant dual root organ culture system sufficiently restored the AM fungal branching response and resulted in functional root colonization leading to production of new viable spores (Sun et al. 2012).

In addition to the different promoting compounds found in the soluble root exudates, also volatile root signals were identified to stimulate AM hyphal growth and branching during the presymbiotic phase and can act synergistically with the soluble exudes compounds (Koske 1982, Bécard & Piché 1989, Gadkar et al. 2003).

2.1.4. Signals Involved in Hyphopodia Formation

After successful attraction of an AM fungus by exuded plant signals, the fungus comes in physical contact with the root surface. To tightly attach to the plant partner, it forms a flattened hyphal structure called hyphopodium. For hyphopodium formation, contact to the root epidermal cell wall is required as demonstrated with isolated cell walls of *D. carota* roots (Nagahashi & Douds 1997). The use of isolated cell walls indicates that neither intact cells nor exuded signals are necessary for hyphopodium formation. Interestingly, hyphopodia formation could not be observed on cell walls of other cell types; it was also found to be absent in some non-hosts such as *Beta vulgaris* (Nagahashi & Douds 1997) and *L. albus* (Gianinazzi-Pearson & Gianinazzi 1992). This clearly indicates that also the step of hyphopodium formation requires specific (host) plant signals. Grafting experiments with *L. albus* and host plants revealed the involvement of the shoot in producing such signal(s) (Gianinazzi-Pearson & Gianinazzi 1992). The recent characterization of *MtRAM1* and *MtRAM2* shed more light on the regulation of AM fungal hyphopodium formation by the plant. The *M. truncatula* mutants *ram1* and *ram2*, whose root exudates lead to normal stimulation of AM fungal germination and hyphal branching, show drastically reduced hyphopodia formation (Gobbato et al. 2012, Wang et al. 2012).

MtRAM1 encodes a GRAS domain transcription factor that regulates *MtRAM2* (Gobbato et al. 2012), which codes for a glycerol-3-phosphate acyl transferase (Wang et al. 2012). Heterologous expression of *MtRAM2* in *Arabidopsis thaliana* and *ram2* mutant analyses indicate that RAM2 functions in the production of cutin monomers belonging to long-chain ω-hydroxy fatty acids (OHFA) and α,ω-dicarboxylic acids (DCA). Roots of *ram2* mutants have next to lower 18:1-DCA and 18:1-OHFA contents reduced levels of 16:0-DCA and 16:0-OHFA, which were also produced in leaves of *MtRAM2*-expressing *A. thaliana* plants (Wang et al. 2012). Application of these C16 cutin monomers, but not of longer-chain lipids, could rescue the *ram2* mutant allowing hyphopodia formation on the surface of *ram2* roots at similar levels than in wild-type plants. For the epidermal surface of the *ram2* mutant, neither modification in the physical structure nor in the wax composition was found compared to the wild type (Wang et al. 2012). These data indicate a signaling (rather than a structural) function of C16:0 cutin monomers on AM fungal hyphopodia formation. Interestingly, also oomycete appressoria formation is regulated by *MtRAM1* and *MtRAM2* indicating a recruitment of the C16:0 aliphatic fatty acid signaling functions by some pathogens (Gobbato et al. 2012, Wang et al. 2012). But different to the interaction with pathogenic oomycetes, both genes are up-regulated upon AM interaction (Wang et al. 2012) suggesting a positive feedback regulation of C16:0 cutin monomer production in the AM symbiosis.

2.1.5. AM Fungal Signals and Signal Perception

Different to appressoria of pathogenic fungi, AM fungi do not use turgor pressure to penetrate the root. Instead, the successful root infection by AM fungi is genetically controlled by the plant. To enable perception by the host and to transcriptionally reprogram the corresponding host cells allowing penetration, also the microsymbiont has to emit specific signals. Several studies showed that diffusible signal(s) from germinating AM fungal spores are able to trigger certain plant responses, including intracellular calcium changes (Navazio et al. 2007, Chabaud et al. 2011), transcriptional activation of symbiosis genes (Kosuta et al. 2003, Navazio et al. 2007, Mukherjee & Ané 2011), and stimulation of lateral root formation (Oláh et al. 2005, Mukherjee & Ané 2011). A response to the diffusible AM fungal signal(s) was found in leguminous and non-leguminous host plants (Mukherjee & Ané 2011), but not in the non-host *A. thaliana* (Navazio et al. 2007). Time course-experiments with spores germinating in the absence of a host indicate that (some) AM fungal signaling molecules, which induce a rapid and transient elevation in the cytosolic calcium concentration in a soybean (*Glycine max*) cell culture, are constitutively released by the fungus (at a lower rate even by un-germinated spores) and thus not necessarily require an induction by plant-derived signals (Navazio et al. 2007). However, to trigger intracellular calcium oscillations, the so-called calcium spiking, in root hair cells of *M. truncatula*, highly branched AM fungal hyphae are required; no calcium changes were observed in root hair cells associated with runner hyphae (Kosuta et al. 2008). This is in line with a recent study that showed strongly elevated exudation of fungal signaling molecules upon spore germination in the presence of the strigolactone analogue GR24 (Genre et al. 2013), which induces hyphal branching.

Interestingly, the plant responses described for the diffusible AM fungal signal(s) are very similar to that elicited by Nod factors, the microbial signals in the LRS. The chemical structure of Nod factors was already elucidated around 25 years ago (Lerouge et al. 1990) as decorated lipochito-oligosaccharides (LCOs), which determine the host-specificity of the rhizobacteria (for overview see D'Haeze & Holsters 2002). Rhizobacteria can produce LCO populations ranging from two to around 60 different individuals, depending on the bacterial strain. Individual rhizobial LCOs structurally differ in the length of the oligosaccharide backbone (four to five β-1,4-linked *N*-acetyl-D-glucosamine residues), the type of the fatty acyl group that is *N*-linked to the non-reducing terminal residue, and/or in the presence and nature of further substituents at the reducing and non-reducing terminal residues that comprise sulphate, methyl, fucosyl, carbamoyl, and acetyl groups. Nod factor production requires expression of rhizobial nodulation (*nod*) genes, which are induced by the transcriptional regulator NodD after interaction with root exuded flavonoids. Rhizobial LCOs are perceived by root hair cells via lysine motif (LysM) domain receptor-like kinases, namely Nod Factor Receptor (NFR)1 and NFR5 in *L. japonicus* (Madsen et al. 2003, Radutoiu et al. 2003, Broghammer et al. 2012) and Nod Factor Perception (NFP) and LYK3 in *M. truncatula* (Ben Amor et al. 2003, Limpens et al. 2003, Arrighi et al. 2006). Nod factor perception rapidly induces host plant responses, such as root hair deformation and curling, transient alkalinization and membrane depolarization, changes in the intracellular calcium concentration, and transcriptional activation of early nodulin genes (for overview see Oldroyd 2013).

Although the existence of corresponding AM fungus-derived signal(s), also referred to as Myc factor, was suggested for a long time, the chemical nature of such an AM fungal factor remained unknown until 2011. In that year, Maillet and co-workers (2011) published the structures of the first identified LCOs exuded by the AM model fungal *R. irregularis*. To purify bioactive LCOs exuded by AM fungi (Myc-LCOs), the authors made use of bioassays that can either sense different non-sulfated LCOs (via root-hair branching in *Vicia sativa*) or sulphated LCOs (via transcriptional activation of the early nodulin gene *ENOD11* in *M. truncatula*). The most active fractions, which also induced lateral root formation, were further purified and LCOs were detected using high-sensitivity, high-resolution mass spectrometry. By that, the authors found a mixture of sulphated and non-sulphated simple LCOs in exudates of mycorrhizal roots and in sterile germinated spore exudates; none were detected in exudates of non-mycorrhizal roots. This confirmed the fungal origin of the identified LCOs. Interestingly, quantities of exuded LCOs as well as the biological activity of AM fungal spore exudates clearly increased upon spore germination indicating a *de novo* synthesis and/or secretion (Maillet et al. 2011). For the identified Myc-LCOs, palmitic acid (C16:0) and oleic acid (C18:1Δ9Z) are the major N-acyl substitutions. To verify and further characterize the biological activity of these Myc-LCOs that were purified with extremely low yield, the authors used bacterial genetic engineering to synthesize individual Myc-LCOs in *Escherichia coli* or in appropriate rhizobial *nod* gene mutants. Corresponding to the isolated natural LCOs, the thus produced sulphated and non-sulphated LCOs were active in *MtENOD11* transcriptional activation and in the root hair deformation assay, respectively. Application of an equal mixture of sulphated and non-sulphated LCOs (each of in a 1:1 mixture with C16:0 and C18:1 N-acyl chains) at a concentration of 10 nM significantly increased number of root infection sites by *R. irregularis* in legumes (*M. truncatula*) and non-legumes (*Tagetes patula*) resulting in enhanced root colonization (Maillet et al. 2011). Root colonization was also increased in root organ cultures of *D. carota*. Thus, the signal processing in plants does not require a shoot. The promoting effect on root colonization was more pronounced for the mixture of non-sulphated and sulfated Myc-LCOs than for the same concentration of pure non-sulphated or pure sulphated LCOs as demonstrated for *T. patula*. However, pure non-sulphated Myc-LCOs stronger increased root colonization than the pure sulphated LCOs. The authors also analyzed the effect of the Myc-LCOs on *M. truncatula* lateral root branching, which is induced by non-sulphated and sulfated Myc-LCOs in a concentration-dependent manner. Here, they found a higher biological activity for sulfated LCOs, which were active down to 0.1 nM or even lower (Maillet et al. 2011). Czaja and co-workers (2012) found also a stronger transcriptional reprogramming in *M. truncatula* roots by sulphated than by non-sulphated Myc-LCOs, which might indicate that a higher number of these genes are involved in the root branching response than in the mycorrhization response. Interestingly, rhizobial LCOs and Myc-LCOs differentially regulated gene expression showing a relatively low number of genes commonly activated by the microbial signals (Czaja et al. 2012). Surprisingly, the mixture of sulphated and non-sulphated Myc-LCOs showed lower transcriptional reprogramming than the pure sulphated Myc-LCOs and also here relatively low overlap was found in the gene activation induced by either sulphated, non-sulphated or mixed Myc-LCOs. Thus, differences in the efficiency and/or specificity in the perception of individual and combined Myc-LCOs are suggested.

Recently, non-LCO AM fungal signals were identified in germinated spore exudates (Genre et al. 2013). Such exudates of *G. margarita* were previously shown to induce a calcium spiking response in epidermal cells (Chabaud et al. 2011), which could be confirmed for exudates of other AM fungi (*Gigaspora rosea*, *R. irregularis*) (Genre et al. 2013). However, the calcium spiking appears irregular and also exhibits cell-to-cell variations, independent on the AM fungal source of exudate. Genre and co-workers (2013) found that a chitinase treatment of the germinated spore exudates completely abolished the spiking response in the root epidermis indicating the chitin-based structure of the signal(s). Application experiments revealed that short-chain chitin oligomers (COs), particularly those of four to five *N*-acetyl-D-glycosamine residues (chitotetraose [CO4] and chitopentaose [CO5]) which could be also detected in germinated spore exudates, efficiently induce non-regular calcium spiking in root organ cultures of *M. truncatula* and of *D. carota* (Genre et al. 2013). When compared to the previously identified non-sulphated and sulphated Myc-LCOs, COs induce the intracellular calcium spiking with higher efficiency. Interestingly, concentrations of short-chain COs, particularly of the highly bioactive CO4 and CO5, strongly increased in AM fungal exudates after strigolactone (GR24) treatment. This was in line with a significantly elevated spiking response. Astonishing, germinated spore exudates of the *M. truncatula* pathogen *Colletotrichum trifolii* were not able to induce calcium spiking although they contained CO4 to a comparable concentration than AM fungal exudates. Even supplementation of the pathogen exudates with CO4 was not sufficient to trigger a calcium response in *M. truncatula* root epidermal cells indicating the presence of additional components inhibiting the CO-induced calcium spiking for this non-AM fungus (Genre et al. 2013).

Corresponding to the LRS, the AM fungal signals are suggested to be perceived by LysM receptor-like kinases (see also Figure 3 in subsection 2.2.). However, so far the Myc signal receptor(s) required for AM formation have not been identified. Recently, the Nod factor receptor gene *NFP* was found to be somehow involved in perception of Myc-LCOs in *M. truncatula*. Here, *NFP* was found to be essential for the Myc-LCO-induced gene expression (Czaja et al. 2012) and to be partly involved in the root branching response elicited by Myc-LCOs (Maillet et al. 2011). However, the root branching phenotype of *nfp* mutants depended on the concentration of applied Myc-LCOs indicating an involvement of other receptor proteins at higher LCO concentrations (Maillet et al. 2011). Moreover, the root branching response to the natural mixture of exuded AM fungal signals and the calcium spiking response to short-chain COs were not found to be affected in *nfp* mutants (Oláh et al. 2005, Mukherjee & Ané 2011, Genre et al. 2013) indicating additional receptor protein(s) and different efficiencies or specificities in the perception of individual and combined AM fungal signals (see also Czaja et al. 2012). Overall, NFP seems not essential for the AM fungal root colonization since *nfp* mutants exhibit a mycorrhization phenotype comparable to that of wild-type plants (Ben Amor et al. 2003, Maillet et al. 2011). Perception of the Myc-LCOs was shown to be also independent of LYK3, which forms a heteromer with NFP for Nod factor perception in *M. truncatula* (Maillet et al. 2011). Thus, other receptor proteins must be involved in perception of AM fungal signals which might be encoded by paralogous genes (Young et al. 2011). The evolution of the LRS was accompanied by a whole-genome duplication event in the papilionoids, the largest subfamily of the Fabaceae, producing genetic redundancy that probably facilitates the evolution of novel, nodulation-specific gene functions without compromising existing functions in mycorrhizal signaling. The paralog of *NFP*, the *LYR1* gene, is highly expressed during AM fungal colonization (Young et al. 2011), but is not required for LRS in which instead *NFP* is transcriptionally activated (Arrighi et al. 2006) indicating functional specialization of the two homologs. In the non-legume *Parasponia andersonii*, that is able to form a rhizobium symbiosis but did not undergo whole genome duplication, the predicted *NFP* ortholog is involved in nodulation and in mycorrhization (Op den Camp et al. 2011). A *LYK3* homolog from *A. thaliana, AtCERK1,* that shows divergence in the activation loop and a three-amino acid (YAQ) motif of the kinase domain from *LYK3* (Nakagawa et al. 2011), is involved in pathogen-elicited defense responses (Zhu et al. 2006, Miya et al. 2007, Willmann et al. 2011, see also Oldroyd 2013). Phylogenetic analyses of *LYK3* homologs from different plant species including the basal legume *Cercis chinensis*, a representative of the Cercideae tribe, indicate a Myc-LCO receptor as common ancestor of the *LYK3* lineage instead of a defense-related receptor kinase (De Mita et al. 2014). After identifying root branching factors and AM fungal signals, the identification of all receptor proteins required for mediating the response to the different AM fungal signaling molecules will be another milestone in elucidating the early signal exchange between the AM partners and the initiation of the host response leading to root infection.

After initial signal perception, further Myc-LCO- and COs-elicited signal transduction requires elements of the common signaling pathway that is conserved for the AM symbiosis and the LRS as outlined below. Compared to non-sulphated and sulphated Myc-LCOs, short-chain COs induce intracellular calcium spiking with higher efficiency. This might indicate undecorated short-chain COs as ancient fungal signals in the AM symbiosis that could (in addition to the mixture of different simple LCOs) contribute to the low host specificity of AM fungi. Elucidating the effects of COs on plant gene expression and on AM fungal root colonization and nodulation will shed more light on the role of COs in the AM symbiosis and its cross-talk with the LRS since different to LCOs COs appear to act NFP-independently. Since some plant responses induced by diffusible AM fungal signals were found to occur independent of the common signaling pathway (Kosuta et al. 2003, Oláh et al. 2005, Mukherjee & Ané 2011) (see also subsection 2.2.), additional signaling pathway(s) involved in AM establishment are suggested, which might be activated either by the identified Myc-LCOs and/or COs or by additional, so far unknown AM fungal signal(s). In addition to such activating signals, the AM model fungus *R. irregularis* DAOM 197198 was found to secret an effector protein, SP7, probably to suppress plant defense responses (Kloppholz et al. 2011). SP7 localizes to the nucleus and interacts with the pathogenesis-related ethylene responsive transcription factor ERF19, apparently promoting the biotrophic status of fungi inside the host. *SP7* transcription is induced in germinated spores and extraradical mycelium by co-incubation with roots and is strongly up-regulated during root colonization (Kloppholz et al. 2011). Transcriptome analyses of this model fungus revealed a whole set of AM fungal genes encoding small secreted proteins that are highly up-regulated in mycorrhizal roots and microdissected arbusculated cells (Tisserant et al. 2013). These small secreted proteins are predicted to act as effectors, similar to SP7, to suppress plant defense mechanisms and/or to manipulate host cell signaling during intracellular accommodation. It can be assumed that a part of those genes is already induced at the presymbiotic phase prior to root infection.

2.2. Activation of a Conserved Signaling Cascade and Transcriptional Reprogramming of the Root Cells

After perception of specific chitin-based microsymbiont signals by (predicted) LysM receptor-like kinases (see above), the signal is further processed by signaling pathways that contain common elements conserved for mycorrhization and nodulation and that finally lead to transcriptional reprogramming of the host cell enabling root infection (see also Figure 3). The common signaling pathway shared in AM and LRS establishment is also called DMI pathway, since it comprises (in the *M. truncatula* nomenclature) the genes *DMI1*, *DMI2*, and *DMI3* (Catoira et al. 2000, Ane et al. 2002). These genes were identified by mutant analyses of *M. truncatula* plants that are unable to form an AM symbiosis and an LRS and were thus called *doesn't make infection* (*dmi*) mutants (Catoira et al. 2000). The main steps of the common signaling cascade in the epidermal root (hair) cells comprise signal perception at the plasma membrane followed by (peri)nuclear calcium spiking that activates a nucleus-localized calcium/calmodulin-dependent protein kinase resulting via activation of transcription factors in specific gene expression. Analyses on *O. sativa*, *S. lycopersicum*, and the tree *Casuarina glauca* demonstrate that this common signaling cascade is not restricted to legumes (Gherbi et al. 2008, Gutjahr et al. 2008, Markmann et al. 2008, Nair & Bhargava 2012). Instead, part of the existing signaling pathways required for successful AM formation appears to be recruited for the later evolved symbioses with nitrogen-fixing bacteria (for review see Markmann & Parniske 2009, Gough & Cullimore 2011, Streng et al. 2011).

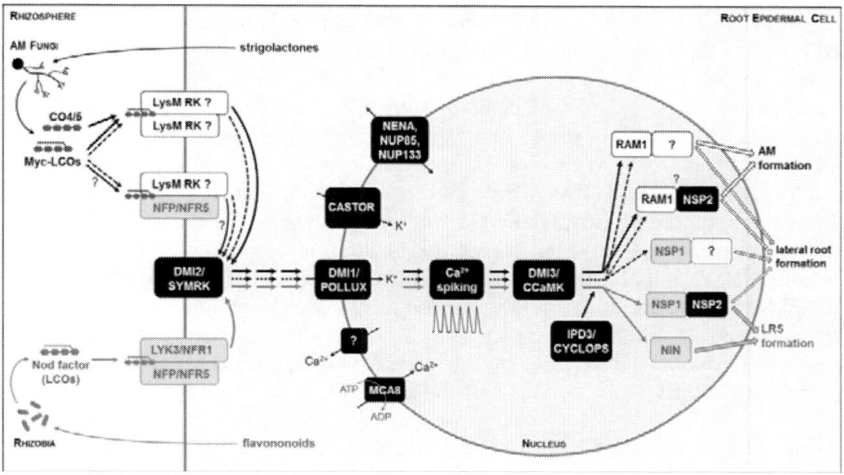

Figure 3. Perception of AM fungal or rhizobial signals and activation of a common signaling cascade resulting in successful root infection. The figure illustrates the elements involved in the establishment of an AM symbiosis or an LRS identified in the model legumes *M. truncatula* and/or *L. japonicus*. The components of the common signaling cascade, also called DMI pathway, are presented as black boxes. Elements specific for the response to AM fungal signals and to Nod factors are illustrated as white and gray boxes, respectively. Ways leading to lateral root formation elicited by Myc signals are indicated with dashed arrows. Additional, so far non-elucidated (DMI-independent) pathways that are activated by AM fungal signals are not illustrated in the schema. Microbial signals are perceived by plasma membrane-bound receptor-like kinases (RKs), which were for the LRS identified as LysM RKs (MtLYK3/LjNFR1 and MtNFP/LjNFR5, Ben Amor et al. 2003, Madsen et al. 2003, Radutoiu et al. 2003) and which are still unknown for the AM symbiosis. Although the Nod factor receptor MtNFP was found to be involved in Myc-LCO-elicited plant responses comprising transcriptional reprogramming (Czaja et al. 2012) and lateral root formation (Maillet et al. 2011), NFP seems not to be essential for mycorrhizal root colonization indicating additional receptor protein(s). Downstream of initial signal perception both in the AM interaction and in the LRS, an additional membrane-bound RK belonging to the LRR class (MtDMI2/LjSYMRK, Catoira et al. 2000, Ane et al. 2002, Stracke et al. 2002) is part of the conserved signaling cascade and involved in producing an unknown secondary messenger. Activation of nuclear membrane-located kation channels, including the potassium channels MtDMI1/LjPOLLUX and LjCASTOR (Ane et al. 2002, Imaizumi-Anraku et al. 2005) as well as still un-identified Ca^{2+} channels and SERCA-type calcium pumps such as MtMCA8 (Capoen et al. 2011), leads to (peri)nuclear oscillations in the calcium concentration, the so-called calcium spiking. Nuclear pore components (NUP85, NUP133, NENA of *L. japonicus*, Kanamori et al. 2006, Saito et al. 2007, Groth et al. 2010) are required for generation of the Ca^{2+} spiking which, however, shows different signatures in the Nod factor and the Myc signal response. The calcium signal is further processed by a calcium/calmodulin-dependent protein kinase (MtDMI3/LjCCaMK[=LjSym15], Lévy et al. 2004, Mitra et

al. 2004, Kistner et al. 2005) which activates, also in interaction with MtIPD3/LjCYCLOPS (Yano et al. 2008, Horváth et al. 2011), symbiosis-specific transcription factors (TFs). For the LRS, MtNIN/LjNIN (Schauser et al. 1999, Marsh et al. 2007) and a GRAS domain TF complex formed by MtNSP1 and MtNSP2 (Hirsch et al. 2009) are essential for transcriptional reprogramming allowing successful root infection and nodule organogenesis. The GRAS TFs might also interact with other TFs (not illustrated). For the pathways activated by AM fungal signals, MtNSP2 was found to be essential for the Myc-LCO-elicited root branching response and to be (partly) involved in mycorrhization (Maillet et al. 2011) – thus representing another common element. MtNSP2 might act together with the GRAS domain transcription factor MtRAM1, which is essential for both AM fungus-induced stimulation of lateral root formation and AM formation (Gobbato et al. 2012). NSP1 was found to be involved in lateral root formation induced also by sulphated Myc-LCOs (next to sulphated rhizobial LCOs) in *M. truncatula* (Maillet et al. 2011).

The LysM receptor-like kinases that perceive the microbial signals are suggested to form a complex with a member of the leucine rich repeat (LRR) kinase family at the plasma membrane. The LysM kinase MtLYK3/LjNFR1 was shown to have functional kinase activity at the cytoplasmic domain (Radutoiu et al. 2003, Arrighi et al. 2006), which might activate the LRR receptor-like kinase. The LRR kinase, that also has a functional kinase domain, is encoded by *DMI2* (also called *NORK*) in *M. truncatula* (Catoira et al. 2000, Ane et al. 2002) and *SYMRK* in *L. japonicus* (Stracke et al. 2002). It probably acts as a co-receptor with the LysM receptors and is as well crucial for the establishment of both, AM symbiosis and LRS. Moreover, a homologous LRR kinase gene identified in *C. glauca* was found to be essential for the development of an actinorhiza (Gherbi et al. 2008, Markmann et al. 2008) – a symbiosis with nitrogen-fixing *Frankia* bacteria that can be formed by eight angiosperm families of the orders Fagales, Cucurbitales and Rosales (Wall 2000). Action of MtDMI2/LjSYMRK is predicted to produce an unknown secondary messenger. MtDMI2 was shown to interact with 3-hydroxy-3-methylglutaryl-CoA reductase (HMGR), more precise with the HMGR1 isoform, which is specifically expressed in roots and which was shown to be essential for root nodule development (Kevei et al. 2007). However, HMGR was recently supposed to play also a role for the AM formation (Kuhn et al. 2010). Since HMGR is a key enzyme in the mevalonate biosynthesis, the secondary messenger might be an isoprenoid compound like cytokinin or a sterol as suggested by Kevei and co-workers (2007). Also, mevalonate is hypothesized as secondary messenger since it was found to activate the following step in the signaling cascade, namely calcium spiking in root hair cells of *M. truncatula* (see Oldroyd 2013 for this information). In addition, phospholipid signaling has been discussed as secondary messenger, at least in Nod factor signaling. It was shown that Nod factor application elevates phospholipid concentration (Den Hartog et al. 2001) and that activity of phospholipase C and/or D is required for induction of Nod factor-elicited root hair curling, calcium spiking, and expression of the early nodulins *ENOD11* and *ENOD12* (Pingret et al. 1998, Den Hartog et al. 2001, Engstrom et al. 2002, Charron et al. 2004).

A secondary messenger is required to translocate the initial signal from the plasma membrane to the root cell nucleus. Here, kation channels become activated that are located in the nuclear envelope and are crucial for the calcium spiking response in the root cell. Among these are MtDMI1/LjPOLLUX and LjCASTOR (Ane et al. 2002, Imaizumi-Anraku et al. 2005). LjCASTOR is an additional homolog from *L. japonicus*, not present in *M. truncatula*, which acts independently of LjPOLLUX but with the same function (Charpentier et al. 2008). These kation channels were previously thought to be associated with the plastids of root cells (Imaizumi-Anraku et al. 2005), but were later identified to be located in the nuclear envelope, at least in case of DMI1 particularly in the inner nuclear membrane (Riely et al. 2007, Charpentier et al. 2008). CASTOR and POLLUX were shown to be particularly permeable for potassium and exhibit only weak calcium permeability (Charpentier et al. 2008). Nevertheless, DMI1/CASTOR and POLLUX are essential for induction of the calcium spiking probably by acting as counter-ion channels compensating the charge changes during each calcium peak by facilitating a potassium influx. Moreover, activation of DMI1/CASTOR and POLLUX likely influences the nuclear membrane potential, which might activate the so far unknown nuclear membrane-located calcium channels (Charpentier et al. 2008).

The calcium spiking occurs in the nucleoplasm and in the nucleus-associated cytoplasm, which is in line with the prediction of the ER as calcium store (Sieberer et al. 2009, Capoen et al. 2011). The ER lumen is contiguous with the lumen of the nuclear envelope. To target the release of the ER calcium store particularly into the nucleus, localization of the involved kation channels to the inner nuclear membrane would be important. Nuclear pores that cross the nuclear envelope allow the transport of molecules between the cytoplasm and the nucleus. In *L. japonicus*, three nuclear pore components, namely NUP85 (Saito et al. 2007), NUP133 (Kanamori et al. 2006), and NENA (Groth et al. 2010), were identified to be crucial for the calcium spiking response and the successful AM fungal and rhizobial entry. These nuclear pore subcomplexes might be involved in selective nuclear import of proteins, including DMI1/POLLUX and CASTOR (see also Groth et al. 2010).

Moreover, nuclear pores are suggested to facilitate the rapid nuclear–cytoplasmic communication after the endosymbiont recognition to enable the activation of the inner nuclear membrane-located kation channels by a secondary messenger. After release of calcium from the internal store, calcium must be retaken up against its concentration gradient to obtain oscillations in the concentration of free calcium. Recently, a calcium pump belonging to the sarco/ER calcium ATPase (SERCA) family, named MCA8, was found to be required for the nuclear calcium signaling during AM and LRS establishment in *M. truncatula* (Capoen et al. 2011). MtMCA8 uses the energy from ATPase hydrolysis to pump calcium back into the lumen of the nuclear envelope and the ER and is located in the inner and outer nuclear membranes and the ER, which would allow efficient reloading of the ER/nuclear envelope calcium store.

Interestingly, rhizobia and AM fungi were found to generate alternative calcium spiking signatures in the pre-infection stage, which might explain the specific responses needed to establish one or the other symbiosis (Kosuta et al. 2008). The calcium signatures are translated by a nucleus-located calcium/calmodulin-dependent serine/threonine protein kinase (CCaMK). It is encoded by *DMI3* in *M. truncatula* (Lévy et al. 2004, Mitra et al. 2004) and by *CCaMK* (formerly named *SYM15*) in *L. japonicus* (Kistner et al. 2005, Tirichine et al. 2006). Activation of the CCaMK by calcium is required – and sufficient – to facilitate transcriptional reprogramming of the root cell and endosymbiont infection (Gleason et al. 2006, Tirichine et al. 2006). The CCaMK can bind calcium directly at three EF hand motifs and via calmodulin that forms a complex with four calcium ions (Ca^{2+}/calmodulin). The enzyme consists of several uniform subunits that possess an autoinhibitory domain which negatively regulates kinase activity. Binding of free calcium by a CCaMK subunit is required for subsequent binding of Ca^{2+}/calmodulin which activates the subunit. Recently it was found, that free calcium not only primes the protein for activation by Ca^{2+}/calmodulin, but also inactivates the protein in the absence of Ca^{2+}/calmodulin-binding by stabilizing the inactive state of the protein (Miller et al. 2013). This different affection of the kinase activity by binding of free calcium and of calcium complexed with calmodulin is suggested to allow discrimination between basal calcium concentrations and those during calcium spiking (Miller et al. 2013). The flexibility in the activation of the CCaMK is predicted to allow a specific decoding of different calcium signatures occurring in response to AM fungal and rhizobial signals (for review see also Singh & Parniske 2012).

Activated CCaMK can induce specific gene expression via activation of transcription factors involved in AM and/or LRS formation. For that, the CCaMK was found to interact with other proteins as phosphorylation substrate, namely MtIPD3/LjCYCLOPS that is essential for both, AM and LRS (Yano et al. 2008, Horváth et al. 2011), and LjCIP73 that was found so far to be crucial for the LRS (Kang et al. 2011). LjCYCLOPS was shown to be a DNA-binding transcriptional activator that forms a homodimer which in interaction with LjCCaMK becomes phosphorylated by the activated kinase (Singh et al. 2014). The phosphorylation results in a conformational change that allows binding and activation of transcription factors by the CCaMK–CYCLOPS complex as demonstrated for NIN (Singh et al. 2014), which is involved in the LRS (Schauser et al. 1999, Marsh et al. 2007). Additional transcription factors required for LRS formation are MtNSP1 (Smit et al. 2005) and MtNSP2 (Oldroyd & Long 2003, Kálo et al. 2005), that form a GRAS domain transcription factor complex (Hirsch et al. 2009). MtNSP1 has DNA-binding domains and can directly interact with promoters of early nodulin genes; in contrast, MtNSP2 appears to lack this ability (Hirsch et al. 2009). The GRAS transcription factors can also interact with additional factors. Recently they were shown to interplay with the ERF transcription factor MtERN1, which is required for LRS and can in addition to MtNSP1 directly bind to promoter regions of early nodulin genes (Cerri et al. 2012). LjNSP2 was shown to interact with the MYB transcription factor LjIPN2 during LRS establishment (Kang et al. 2014).

In a recent study, the GRAS transcription factor MtNSP2 was identified as another common element involved in the response to rhizobial and AM fungal signals. Maillet and co-workers (2011) demonstrated in *M. truncatula* mutant analyses that *NSP2* is essential for the Myc-LCO-induced root branching response and also involved in root mycorrhization. However, AM fungal root colonization was significantly reduced but not abolished in the *nsp2-2* mutant (Maillet et al. 2011) indicating additional factors mediating the mycorrhization response. In contrast to *MtNSP2*, *MtNSP1* is not involved in formation of an AM symbiosis; instead it is required for lateral root formation induced by sulphated Myc-LCOs, but not by non-sulphated Myc-LCOs (Maillet et al. 2011). This partial involvement of *NSP1* in Myc signal-elicited root branching of *M. truncatula* is probably due to structural similarity with sulphated Nod factors of *Sinorhizobium meliloti* which are bioactive in this legume species. The NSP1-independent root branching response to non-sulphated Myc-LCOs is suggested to be mediated by a pathway downstream of DMI3 that is different to the Nod factor signaling pathway (Maillet et al. 2011). To enable action of the common transcription factor NSP2 in processes not dependent on NSP1, it must form a complex with other transcriptional regulators that have a direct DNA-binding capacity. One candidate might be the GRAS transcription factor MtRAM1, which was shown to interact with MtNSP2 *in vitro*. MtRAM1 is essential for both, Myc-LCO-induced stimulation of lateral root formation and for the formation of AM fungal hyphopodia (Gobbato et al. 2012). It can directly bind to the promoter and elicit transcription of the mycorrhiza-inducible gene *MtRAM2* that codes for a glycerol-3-phosphate acyl transferase involved in cutin biosynthesis (Gobbato et al. 2013) (see also subsection 2.1.4.).

Interestingly, differences were found in the elements upstream and downstream of the DMI pathway that are involved in the mycorrhization response and in the root branching response elicited by AM fungal signals in *M. truncatula*. The Nod factor receptor NFP is partly involved in the stimulation of lateral root formation by sulphated as well as non-sulphated Myc-LCOs, but has no function in the mycorrhization response (Maillet et al. 2011). Downstream of the calcium/calmodulin-dependent protein kinase DMI3, different roles in Myc signal-induced root branching and fungal root colonization were found for the transcription factors NSP1 and NSP2 as described above. The different dependencies of the Myc signal-elicited root branching response and the mycorrhization response on specific pathway components indicate multiple DMI-dependent signaling pathways involved in the complex Myc signal response. The individual pathways might be also activated to a different degree by individual AM fungal signals (sulphated Myc-LCOs, non-sulphated Myc-LCOs, short-chain COs, and putative additional, so far un-identified signals). In addition, in leguminous plants, whole-genome duplication and partial recruitment of the pathway(s) for the evolution of the LRS, which was also accompanied by increased functional specialization, might have contributed to the complexity found in model legumes. Stimulation of lateral root branching is a response common to Myc signals and to Nod factors. This is in line with the higher grade of conservation of the therein involved pathways as compared to the pathways resulting in root colonization with the individual endosymbionts.

Moreover, some responses to AM fungal signals are (additionally) mediated by DMI-independent pathways. For example, lateral root formation induced by non-identified diffusible AM fungal factor(s) in *M. truncatula* were found to require *DMI1* and *DMI2*, but not *DMI3* (Oláh et al. 2005, Mukherjee & Ané 2011), whereas in *O. sativa* also the *DMI2* ortholog *OsCASTOR* and *OsPOLLUX* were not required (Mukherjee & Ané 2011).

Transcriptional activation of the early nodulin gene *MtENOD11* by unidentified diffusible AM fungal signals (probably a slowly diffusible or relatively unstable factor of low molecular mass) was found to be independent of *MtDMI1*, *MtDMI2*, and *MtDMI3* (Kosuta et al. 2003). In contrast, *MtENOD11* expression in epidermal cells in contact with AM fungal hyphopodia requires at least part of the DMI pathway as demonstrated with the *dmi2-2* mutant (Chabaud et al. 2002). In *O. sativa*, induction of two early AM response genes was found to occur independent of *OsCASTOR* and *OsPOLLUX* and of the *DMI3* ortholog *OsCCaMK*, whereas expression of other AM-inducible genes required these elements (Gutjahr et al. 2008), All this points to the high complexity of the Myc signaling pathways which might also enable certain flexibility in the response to the natural mixture of exuded AM fungal signals.

2.3. Root Infection by AM Fungi and Intracellular Accommodation

2.3.1. Host Cell Infection

The AM fungal infection of the root is a process supported by the plant partner. AM fungi attach to the root surface with a hyphopodium which does not enable penetration of the root cell wall by turgor pressure as with an appressorium of pathogenic fungi. Besides, AM fungi lack genes encoding plant cell wall-degrading enzymes (Tisserant et al. 2013). Instead, successful root penetration and colonization by AM fungi are genetically controlled by the plant. Intracellular epidermal penetration by AM fungi can occur at the outer epidermal cell wall as shown e.g. for *M. truncatula* and *D. carota* (Genre et al. 2005, 2008) and/or at the anticlinal wall between adjacent epidermal cells, as described e.g. for *L. japonicus* (Bonfante et al. 2000) as well as for *M. truncatula* (Genre et al. 2005). Also, an intercellular epidermal passage can take place followed by an intracellular penetration of the exodermis or of outer cortical cells as found e.g. in *L. japonicus* (Bonfante et al. 2000, Demchenko et al. 2004), *M. truncatula* and *D. carota* (Genre et al. 2008). However, in all cases root infection comprises at least one step of intracellular infection prior colonization of the inner cortex, and it is always dependent on specific symbiosis genes including the elements of the DMI pathway. Next to common symbiosis genes (see above), additional genes are required for successful AM fungal entry into a host root. Recently, a *M. truncatula* gene encoding a predicted membrane steroid-binding protein named MtMSBP1 was identified to be essential for AM establishment, suggesting the involvement of steroid signals (Kuhn et al. 2010). It might be possible, that MtMSBP1 is involved in mediating the response to the so far unidentified secondary messenger after initial Myc signal perception involving HMGR (see subsection 2.2.). MtMSBP1 is transcriptionally activated early in the AM interaction requiring DMI2 and was found to be present in the ER network including the nuclear envelope. RNAi-mediated gene suppression resulted in a frequent abort of epidermal penetration by septation of the penetrating hypha and strongly impaired arbuscule formation (Kuhn et al. 2010). A similar phenotype with early arrest of invading hyphae and strong defects in root colonization was found for *d3* mutants of *O. sativa* (Yoshida et al. 2012). *D3* encodes an F-box protein that is essential for the response to strigolactones. However, another strigolactone insensitive rice mutant (*d14*), with defects in a predicted α/β-hydrolase gene, showed a wild-type mycorrhization phenotype (Yoshida et al. 2012). Interestingly, strigolactone insensitive mutants can differ in their responsiveness to strigolactone and non-plant-derived compounds and in their complementation by G24 (Hu et al. 2010, Waters et al. 2012). Accordingly, the D3 F-box protein is predicted to be (in addition) involved in a signaling pathway independent on strigolactones that is required for root infection after hyphopodium contact (Yoshida et al. 2012). Other genes are found to act later in AM formation and/or AM functioning as outlined below.

Intracellular AM fungal infection and accommodation requires separation from the cytoplasm of the living plant cell by a plant-derived perifungal membrane (in arbusculated cells also called periarbuscular membrane) and an apoplastic compartment, the symbiotic interface. The process of intracellular AM fungal penetration and the construction of the symbiotic interface is preceded by the development of a plant-derived tunnel-like structure as microscopically uncovered in *M. truncatula* and *D. carota* roots by Genre and co-workers (2005, 2008) (Figure 4). By using fluorescent labeling of the plant cytoskeleton and the endoplasmic reticulum (ER) and confocal laser scanning microscopy, the authors revealed that in response to AM fungal hyphopodium formation, a nucleus-directed cytoskeletal/ER apparatus, the so called prepenetration apparatus (PPA), assembles in the corresponding epidermal cell (Figure 4a,b). The PPA develops within few hours after hyphal contact requiring, similar to the infection thread formation in the LRS, the DMI signaling pathway including calcium spiking and is preceded by transcriptionally activation of *ENOD11* in *M. truncatula*. It is formed during a two-step nuclear migration (see Figure 4 for details) and comprises dense ER cisternae surrounded by high-density arrays of microtubules and microfilament bundles formed within a cytoplasmic column traversing the epidermal cell from the site below hyphopodium contact to the opposite side bordering the cortex. Along this PPA, the fungus grows across the lumen (Figure 4c,d). This takes place after invagination of the plant plasma membrane initiating the formation of the perifungal membrane and the symbiotic interface in which the PPA plays a key role. The PPA was found to be a transient structure that dismantles after fungal entry and formation of the symbiotic interface. Interestingly, PPA formation is specific to the (compatible) symbiotic fungal interaction, whereas an abiotic mechanical stimulus and pathogenic fungi are sufficient to trigger nucleus migration and cytoplasmic aggregation, respectively (Genre et al. 2009).

The PPA also operates during intracellular penetration of the outer cortex and colonization of the inner cortex by AM fungi, both in *Arum*-type and *Paris*-type mycorrhizas as demonstrated for the interaction of *Gigaspora gigantea* with *M. truncatula* and *D. carota*, respectively (Genre et al. 2005, 2008) (Figure 4c-h). Outer cortical cells located below the infected epidermal cell respond even before physical contact with the invading AM fungal hypha (Genre et al. 2008, Sieberer et al. 2012). Similar to epidermal entry, outer cortical cell invasion by AM fungi is preceded by (peri)nuclear calcium spiking in the corresponding and in adjacent plant cells as demonstrated for *M. truncatula* (Sieberer et al. 2012). The calcium signal starts at low frequency, followed by nuclear migration below the epidermal nucleus/PPA, and precedes AM fungal invasion of an outer cortex cell as a high-frequency spiking (Figure 4c,d). The calcium signal ends after successful cell infection. The switch from low- to high-frequency calcium spiking is supposed to determine the switch from reversible processes (including PPA formation) to irreversible cell infection processes initiated by the plant (like *de novo* interface synthesis) (Sieberer et al. 2012).

In the inner cortex, PPA formation and structure correspond to the respective colonization strategy of *Arum*- and *Paris*-type mycorrhizas (Genre et al. 2008). In *M. truncatula* roots establishing an *Arum*-type AM, inner cortical cells form a PPA after physical contact with an intercellular fungal hypha; in the *Paris* type in *D. carota*, linear polarized PPAs were found within several inner cortical cell files either in an acropetal or basipetal orientation, according to the direction of fungal growth. In both, after infection of an inner cortical cell along the PPA, formation of arbuscular branches is preceded by the development of prebranching cytoplasmic aggregations along the invaded hypha. Interestingly, prior to infection of the inner cortical cells, nuclear enlargement was found which was associated with chromatin decondensation and large, highly organized nucleoli (Genre et al. 2008), both suggesting intense transcriptional activity. That is in line with transcriptional reprogramming prior and during arbuscule formation of plant cells (see for example Gomez et al. 2009, Guether et al. 2009a, Hogekamp et al. 2011, Gaude et al. 2012, Hogekamp & Küster 2013), which is probably, at least partly, mediated by CYCLOPS/IPD3 since gene mutations can particularly affect arbuscule formation (Yano et al. 2008, Horváth et al. 2011). A high proportion of the regulated genes can be linked to functions in protein turnover, transport processes as well as membrane proliferation and cell wall synthesis and remodeling.

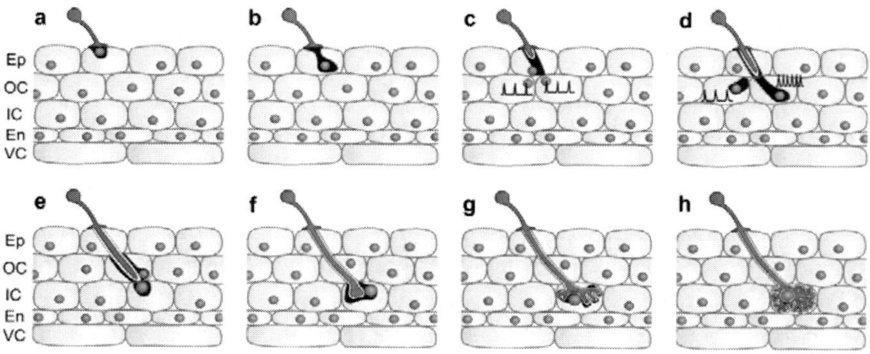

Figure 4. AM fungal root infection and intracellular accommodation of the endosymbiont is guided by a prepenetration apparatus (PPA). Initiation of PPA formation (Genre et al. 2005, 2008) requires the DMI pathway including Ca^{2+} spiking. After AM fungal hyphopodium formation on the root surface, the epidermal cell nucleus rapidly translocates beneath the site of hyphopodium contact accompanied by cytoskeleton rearrangements and by a cytoplasmic aggregation and formation of a ring-like ER structure below the hyphopodium contact site (a). Then, the nucleus migrates across the cell lumen (often in a diagonally way) to the opposite, cortex-oriented side of the cell forming by this the tubular-structured PPA comprising ER, microtubules, and actin filaments (b). The PPA is involved in the synthesis of the apoplastic compartment required for intracellular accommodation of the invading fungus. Along the PPA, the fungus penetrates the epidermal cell and grows across the lumen (c). After the hypha traversed the epidermal cell, the cell nucleus appears to migrate back to the cell periphery (d) and the PPA starts to degenerate (e). AM fungal ingrowth into the root cortex is preceded by calcium spiking starting at low frequency and by nuclear movement towards the terminal end of the epidermal PPA in the corresponding and adjacent outer cortex cells (c), which both start to initiate transcellular PPA formation (d) (Sieberer et al. 2012). The cortex cell that will be penetrated switches to high-frequency calcium spiking prior to cell infection which is preceded by membrane invagination to synthesize the symbiotic interface (d). The calcium signal ends after successful cell infection and the hypha passes through the

outer cortical cell; the PPA in the adjacent cell rapidly disassembled (e). Intracellular access to the inner cortex is also preceded by PPA formation and here also by nucleus enlargement (Genre et al. 2008) (e). During arbuscule development, the PPA functions in the formation of hyphal branches via prebranching cytoplasmic aggregations and in the formation of the periarbuscular membrane (d). In arbusculated cells, the enlarged nucleus is centrally positioned (e) and a layer of ER-rich cytoplasm surrounds the arbuscule branches; the central vacuole is partly fragmented (Pumplin & Harrison 2009). Ep: epidermis; OC: outer cortex; IC: inner cortex; En: endodermis; VC: vascular cylinder.

Particularly the arbusculated cells undergo extensive cellular reorganization events, including periarbuscular membrane biogenesis and development of the symbiotic interface (see below), that are required for functional AM symbiosis. Arbuscule formation is also accompanied by changes in the plastid and mitochondria abundance and structure. In the inner cortical PPA, polymorphic mitochondria and chromoplast-like plastids, both often elongated, were detected (Genre et al. 2008). Upon arbuscule development, extensive proliferation of plastids and mitochondria and formation of network-like organelle structures surrounding the arbuscule were shown indicating the increased metabolic activity of arbusculated cells (Fester, Hans et al. 2004, Lohse et al. 2005, 2006). The stimulated metabolism is necessary to drive the plant cellular rearrangements that are essential for intracellular arbuscule development. Moreover, activation of the apocarotenoid metabolism, that goes along with accumulation of C_{13} cyclohexenone and C_{14} mycorradicin apocarotenoids in mycorrhizal roots (Strack & Fester 2006, Schliemann et al. 2008), appears to be involved in producing signals affecting the arbuscule life cycle and mycorrhizal functionality at later stages (Walter et al. 2007). Knock-downs of genes encoding an element of the non-mevalonate pathway and an carotenoid-cleavage protein result both in a higher proportion of senescing and dead arbuscules (Floß et al. 2008a, 2008b). Although arbuscule development is known to be facilitated by effective nutrient exchange (see below), the signaling pathways underlying cortex cell infection and arbuscule development are still less understood.

For several phytohormones a role in AM establishment and functioning has been suggested or demonstrated; however, data are sometimes contradictory and complex phytohormone cross-talks can occur as reviewed in more detail elsewhere (Ludwig-Müller & Güther 2007, Hause & Schaarschmidt 2009, Ludwig-Müller 2010, Foo et al. 2013a). For example, in *S. lycopersicum*, abscisic acid (ABA) is involved in specific transcriptional reprograming and proper arbuscule formation and functioning as demonstrated with ABA-deficient mutant plants (Herrera-Medina et al. 2007, Garrido et al. 2010, Martín Rodriguez et al. 2010). ABA is also suggested to regulate strigolactone biosynthesis (López-Ráez et al. 2010a). Moreover, ABA deficiency in *S. lycopersium* leads via cross-talking to ethylene accumulation (Martín Rodriguez et al. 2010), which can reduce root colonization and is partly involved in mediating the AM phenotype of ABA-deficient mutants (de Los Santos et al. 2011, Martín-Rodríguez et al. 2011). In wild-type tomato, ethylene accumulation is reduced upon mycorrhization (López-Ráez et al. 2010b). In contrast to ethylene, the oxylipin pathway/jasmonic acid (JA) biosynthesis was found to be induced upon root colonization in different plant species (Hause & Schaarschmidt 2009, López-Ráez et al. 2010b). Suppression of JA biosynthesis in *M. truncatula* roots resulted in slower root colonization and reduced arbuscule functioning, but no obvious morphological changes in arbuscule structure (Isayenkov et al. 2005). An important function of gibberellins for arbuscule formation was recently shown by identification of two DELLA proteins, MtDELLA1 and MtDELLA2, negative regulators of gibberellic acid (GA) signaling that appear to act independently of the DMI signaling pathway (Floß et al. 2013). Double mutation of both genes led to strongly reduced arbuscule formation and reduced *NSP1/NSP2* transcript accumulation indicating a repressing effect of GA on functional AM formation. This is in line with a study on a gibberellin-deficient pea (*Pisum sativum*) mutant that was found to exhibit a higher colonization rate and an increased arbuscule number (Foo et al. 2013a).

Mycorrhizal roots can also show elevated auxin levels; for *M. truncatula* endogenous root auxin concentrations were found to peak prior/during root infection and at late stages of AM interaction (Ludwig-Müller & Güther 2007). An auxin rise might force strigolactone signaling as an *P. sativum* auxin mutant showed reduced strigolactone exudation (Foo 2013) and the *P. hybrida* strigolactone exporter PhPDR1 was found to be transcriptionally activated by the auxin analogue 1-naphthaleneacetic acid (NAA) (Kretzschmar et al. 2012). The ABC transporter PhPDR1 is strongly expressed in and around colonized cells and mutation resulted in a delayed colonization similar to that found in *P. hybrida* strigolactone biosynthesis mutant (Kretzschmar et al. 2012). However, the intracellular AM fungal structures in these mutant plants were of regular morphology and no aberrations could be detected, which corresponds to results on a strigolactone biosynthesis mutant of rice (Gutjahr et al. 2012). These data and results of application experiments (Foo 2013) suggest that strigolactones can also stimulate intracellular fungal growth and arbuscule formation, but that they are not essential for root colonization and arbuscule branching.

Recently, two half-size ABC transporters, STR(1) and STR2, were identified to be crucial for full arbuscule development in *M. truncatula* and *O. sativa* (Zhang et al. 2010, Gutjahr et al. 2012). MtSTR and MtSTR2 can form a heterodimer that is located in the periarbuscular membrane (Zhang et al. 2010). The STR(1)/STR2 transporter is predicted to export its substrate to the symbiotic interface surrounding the arbuscule (Zhang et al. 2010); however, so far their substrate is still unknown, but probably different to strigolactones (Gutjahr et al. 2012). A putative member of the APETALA 2/ethylene-responsive element binding protein (AP2-EREBP) transcription factor family, whose members are involved in hormone, sugar and redox signaling (Dietz et al. 2010), was found to be specifically expressed in arbusculated cells and artificial microRNA-mediated gene silencing demonstrated the requirement of MtErf1 for full arbuscule development (Devers et al. 2013). Overexpression of the early symbiosis gene *enod40*, a short open reading frame-containing RNA, was found to stimulate AM fungal root colonization and arbuscule formation (Staehelin et al. 2001). *Enod40* is involved in cytoplasmic relocalization of nuclear proteins (Campalans et al. 2004). Moreover, apoplastic subtilisin proteases, SbtM1 and SbtM3 of *L. japonicus*, are somehow involved in arbuscule formation as gene silencing strongly reduced the number of arbuscules and in addition lowered the number of internal hyphae (Takeda et al. 2009). The SbtM1 protein appears to localize in the apoplast around infected cells and particularly in the symbiotic interface around intracellular fungal hyphae and arbuscules. Additional genes known to be involved in intracellular AM fungal accommodation and arbuscule development encode proteins associated with exocytotic vesicle trafficking and symbiotic phosphate transporters and are described below. To generally enable intracellular fungal development, formation of the symbiotic interface is required, representing a key step in regulating root colonization and arbuscule formation.

2.3.2. Formation of the Symbiotic Interface

The formation of the PPA and the construction of the symbiotic interface and the perifungal membrane are associated with substantial membrane and cell wall dynamics, particularly during arbuscule formation (for review see Balestrini & Bonfante 2014). Such membrane/cell wall dynamics are not only crucial for functional AM establishment, but also functions during other biotrophic interactions, such as the ectomycorrhizal symbiosis and LRS as well as interactions with plant-parasitic nematodes or with biotrophic microbial pathogens (for review see Brewin 2004, Balestrini & Bonfante 2014, Bellincampi et al. 2014, Rodiuc et al. 2014). In the AM symbiosis, synthesis of the perifungal/periarbuscular membrane in the PPA involves extensive exocytotic activities, with major participation of the Golgi apparatus often associated with microtubules and fused into extensive tubular-vesicular trans-Golgi networks (Genre et al. 2012). Golgi stacks particularly accumulate ahead of the tips of developing intracellular hyphae and around branches of developing arbuscules (Pumplin & Harrison 2009, Genre et al. 2012) accompanied by accumulation of secretory vesicles and of the exocyst complex near the developing perifungal membrane (Genre et al. 2012). Proteins associated with vesicle trafficking have recently been identified to have a key function in AM formation. During exocytosis, the fusion of vesicles with plasma membranes is mediated by SNARE proteins – in plants by SNARE proteins of the exocytotic vesicle-associated membrane protein (VAMP)72 family. In *M. truncatula*, VAMP721d and VAMP721e are transcriptionally activated upon mycorrhization and present in secretory vesicles and the newly formed perifungal/periarbuscular membrane near the growing hyphal tip, and in case of VAMP721e also on the side of the penetrating hypha and near fine arbuscular branches (Genre et al. 2012, Ivanov et al. 2012). Simultaneous knock-down of both genes was found to block arbuscule formation (Ivanov et al. 2012).

Another protein crucial for AM formation in *M. truncatula* is Vapyrin, a novel protein predicted as VAMP-associated protein (VAP) that is absent in *A. thaliana* (Pumplin et al. 2010). In *P. hybrida*, a Vapyrin homolog is encoded by *RAM1* (Reddy et al. 2007, Feddermann et al. 2010). Vapyrin is essential for intracellular invasion of AM fungal hyphae into root cells. Gene knock-down or mutation strongly impaired epidermal penetration after hyphopodium formation and disabled arbuscule formation (Reddy et al. 2007, Feddermann et al. 2010, Pumplin et al. 2010). In wild-type *M. truncatula* and *P. hybrida* roots, *Vapyrin* is transcriptionally activated during the epidermal infection process and during arbuscule formation; petunia *Vapyrin* transcripts particularly accumulate in cells containing finely branched hyphae (Feddermann et al. 2010, Pumplin et al. 2010). MtVapyrin accumulation precedes AM fungal invasion and was found below hyphopodia in the cytoplasmic aggregation that occur during PPA development and in arbusculated and adjacent cells (Pumplin et al. 2010). In infected cells, Vapyrin protein accumulates in small mobile puncta, probably vesicles (Feddermann et al. 2010, Pumplin et al. 2010).

For *M. truncatula*, all three proteins associated with exocytotic vesicle trafficking were also found to be recruited by the LRS and are essential for rhizobia infection. MtVapyrin and MtVAMP721d/e function in formation of infection threads and symbiosomes, respectively, both represent processes connected to membrane biogenesis (Murray et al. 2011, Ivanov et al. 2012). In addition to the exocytotic activity preceding and accompanying intracellular AM fungal accommodation, presence of multivesicular bodies in colonized cells indicate also an involvement of endocytotic processes (Genre et al. 2008). The detected endosomes might function in recycling plasma membrane proteins back to the membrane or in signaling during interface formation.

The newly developed symbiotic interface between the AM fungal membrane and the plant-derived perifungal membrane contains cell wall polysaccharides and cell wall-associated proteins, such as hydroxyproline-rich glycoproteins (HRGPs) including arabinogalactan proteins (AGPs) and expansins, mostly of host origin (see e.g. Bonfante-Fasolo et al. 1990, 1991, Balestrini et al. 1996, 2005, Schultz & Harrison 2008). In the *M. truncatula–R. irregularis* interaction, additionally to an AM-inducible plant *AGP* gene, AM fungal derived AGPs were found to be expressed that are suggested to assist the interface formation through self-assembly and interactions with plant cell surfaces (Schultz & Harrison 2008). One predicted plant cell wall repetitive proline-rich protein is encoded by the early nodulin gene *MtENOD11*, whose expression precedes and accompanies PPA formation in epidermal and cortex cells (Genre et al. 2005, 2008) and arbuscules development (Journet et al. 2001) and which also appear to be closely associated with infection thread formation in the LRS (Journet et al. 2001).

Expansins, proteins involved in cell wall-loosening during cell wall extensibility (Cosgrove 2000), are suggested to facilitate cell wall plasticity of the symbiotic interface. This is particularly necessary in arbusculated cells since the interface morphology changes along the arbuscule structure and during the life cycle of the arbuscule (Pumplin & Harrison 2009, Balestrini & Bonfante 2014). Expansin(-like) genes were found to be induced during early and later stages of an AM interaction including epidermal cells with hyphopodium contact and arbusculated cells, respectively (Weidmann et al. 2004, Siciliano et al. 2007, Dermatsev et al. 2010). Knock-down of an AM-induced expansin-like gene in *S. lycopersicum* lead to lower AM fungal infection rates and reduced arbuscule expansion (Dermatsev et al. 2010). Immuno-analyses revealed that α-expansin proteins accumulate in the symbiotic interface and in the cell wall of arbusculated cells, which exhibit increased cell size and cell wall thickness (Balestrini et al. 2005). The interface components indicate that the newly formed perifungal membrane maintained its ability to synthesize and secrete cell wall material. However, compared to the plasma membrane from which it originates from, the periarbuscular membrane exhibit a different membrane identity particularly characterized by a high density of different nutrient transport systems.

2.3.3. Periarbuscular Membrane and Mineral Nutrient Uptake

To provide for the improved delivery of mineral nutrients trough the AM interaction, plants express specific transporter proteins upon AM interaction that are particularly localized to the periarbuscular membrane. Among those are AM-induced phosphate transporters whose expression upon mycorrhization is usually accompanied with a down-regulation of the non-symbiotic transport mechanism for inorganic phosphate (Pi) (for review see Bucher 2007, Javot et al. 2007b, Smith et al. 2011, Yang & Paszkowski 2011). AM-inducible phosphate transporters are either exclusively expressed in response to an AM symbiosis or up-regulated upon mycorrhization with basal levels in non-mycorrhizal roots. The symbiotic phosphate uptake in plants can comprise two non-orthologous transport systems involving Pi transporters of the Pht1 phosphate transporter subfamilies I and III. Mycorrhiza-specific Pht1 subfamily I transporters were first identified for *M. truncatula* (the low affinity Pi transporter MtPT4, Harrison et al. 2002) and *O. sativa* (the high affinity Pi transporter OsPT11, Paszkowski et al. 2002). Since then, homologs were found in several plant species including Solanaceae species (Nagy et al. 2005), legumes (Maeda et al. 2006, Tamura et al. 2012), and monocots (Glassop et al. 2005, Nagy et al. 2006). AM-inducible members of the Pht1 subfamily III include the high affinity phosphate transporter gene *StPT3* from potato (*Solanum tuberosum*) that is strongly induced upon mycorrhization and which was the first identified AM-inducible plant phosphate transporter (Rausch et al. 2001). So far, homologs have been identified for *S. lycopersicum* (Nagy et al. 2005) and *L. japonicus* (Maeda et al. 2006); however, the monocot *O. sativa* is lacking a corresponding homolog (Paszkowski et al. 2002) suggesting an evolution of this second symbiotic Pi transport system in dicots only.

Promoter–GUS fusions, *in situ* hybridizations, and laser microdissection revealed for several of those AM-inducible phosphate transporters an expression specific to arbusculated cells (Rausch et al. 2001, Harrison et al. 2002, Glassop et al. 2005, Nagy et al. 2005, Balestrini et al. 2007). Immunolabeling showed that MtPT4 is localized in the periarbuscular membrane in the region around the hyphal branches, no protein accumulation was found around the trunk hypha of arbuscules or in the plasma membrane of arbusculated cells (Harrison et al. 2002, Pumplin & Harrison 2009). Moreover, presence of MtPT4 is determined by the life cycle of the arbuscule and the protein appears to be degraded upon arbuscule collapse. In *S. tuberosum* that exhibit *Paris* or intermediate AM morphology types, *StPT3* is expressed in cortex cells harboring arbusculate coils or thick-coiled hyphae lacking arbuscule branches (Karandashov et al. 2004), both are suggested to function in the phosphate release in *Paris* type AM (Smith & Smith 1997, van Aarle & Olsson 2003). Thus, induction of such symbiotic phosphate transporters appear to be specific to perifungal membranes surrounding those AM fungal structures that represent the fungal organs active in mineral nutrient exchange.

A crucial role of the symbiotic phosphate transport for successful AM establishment was demonstrated by gene suppression and mutation of AM-inducible Pi transporters. MtPT4 (Pht1 subfamily I member) and LjPT3 (Pht1 subfamily III member) were both found to be involved in the full development and/or stability of fungal structures, particularly arbuscules. Strikingly, *LjPT3* RNAi knock-down roots not only showed a reduced number of arbuscules, but here also defense responses seem to be activated upon mycorrhization since co-inoculation with rhizobia resulted in necrotic nodules whereas a single rhizobia-inoculation lead to healthy nodule formation (Maeda et al. 2006). In *MtPT4* RNAi plants, arbuscules were characterized by a strongly reduced number and by higher rate of collapsing and degenerating arbuscules over time accompanied with septation of intercellular hyphae indicating death in AM fungi (Javot et al. 2007a). The dependency of AM formation on symbiotic phosphate transfer was confirmed with an *mtpt4* mutant, that show a similar intraradical colonization phenotype. Here, the AM fungus was not able to develop an extraradical mycelium (Javot et al. 2007a). This clearly indicates the regulation of the AM symbiosis via its functionality.

The transcriptional activation of AM-responsive Pi transporter genes, namely *StPT3* and *StPT4* of *S. tuberosum* and *SlPT4* of *S. lycopersicum*, was recently shown to be specifically elicited by the lysophospholipid lysophosphatidylcholine (LCP) (Drissner et al. 2007). This compound was found in the phospholipid extract of mycorrhizal roots. Bioactivity was demonstrated for phospholipid extracts of mycorrhizal roots of different plant species, whereas extracts of non-mycorrhizal roots as well as of extraradical AM fungal material were inactive (Drissner et al. 2007). LCP can be produced by hydrolyzation of the phospholipid phosphatidylcholine, a membrane component which showed no bioactivity, by phospholipase A_2 (Drissner et al. 2007). LCP is highly mobile within intact cells and is suggested to act as a cytoplasmic messenger in arbuscule-/coil-containing cells representing a late signal in the AM symbiosis. So far it is unknown if the LCP in the root lipid fraction of mycorrhizal roots is of host plant or AM fungal origin (Drissner et al. 2007). Interestingly, LCP-mediated induction of AM-inducible Pi transporter genes can be suppressed by a high plant P status (Nagy et al. 2009). This P-dependency might partly explain the repressory effect of LCP on gene expression of AM-inducible Pi transporters that was detected in *P. hybrida* plants 40 days after mycorrhizal inoculation (Tan et al. 2012). However, AM inducible Pi transporter genes identified in *P. hybrida* appear to exhibit different responsiveness to LCP and no expression of the AM-specific *PhPT4* gene was found upon LCP treatment in non-mycorrhizal plants (Tan et al. 2012). Thus, further experiments will help to elucidate the role of LCP and/or other phospholipids during the AM interaction in different plant species.

Plant Pht1 transporters, comprising the so far identified AM-inducible Pi transporters, are Pi:H^+ symporter (Nussaume et al. 2011). Accordingly, next to the perifungal membrane-located Pht1 transporter(s), high abundance of H^+-ATPases was detected in the periarbuscular membrane (Gianinazzi-Pearson et al. 2000) accompanied by an increased ATPase activity particularly around the finer arbuscule branches (Marx, Gianinazzi-Pearson et al. 1991). *De novo* synthesis of H^+-ATPases upon mycorrhization appears to be specific for arbusculated cells as indicated by transcript analyses (Gianinazzi-Pearson et al. 2000, Krajinski et al. 2002). The high H^+-ATPase activity in the periarbuscular membrane goes along with a highly acid nature of the symbiotic interface (Guttenberger 2000). The pH was found to be markedly lower than the vacuolar pH (Guttenberger 2000); measurement with pH-sensitive fluorochromes indicates a pH value around 4 (Smith et al. 2001). This corresponds to the pH optimum found for MtPT4 activity in a heterologous expression system (Harrison et al. 2002). Moreover, an apoplastic pH below 5.5 also increases the activity of cell wall-loosening enzymes showing high activity at pH values ranging from 3 to 5 with a maximum between 3.5 and 4.5 (McQueen-Mason et al. 1992, Cosgrove 2000).

In addition to phosphate transporters, the periarbuscular membrane is predicted to contain several other AM-inducible macro- and micronutrient transporters. Different AM-inducible genes encoding putative ammonium, potassium, sulfate and micronutrient (zinc, copper, manganese) transporters were found to be specifically induced in arbusculated cells (Gomez et al. 2009, Guether et al. 2009a, 2009b, Kobae et al. 2010, Hogekamp et al. 2011, Gaude et al. 2012, Koegel et al. 2013); for the ammonium transporters GmAMT4.1 from *G. max* and SbAMT3;1 from *Sorghum bicolor* a localization in the branch domain of periarbuscular membranes has been demonstrated (Kobae et al. 2010, Koegel et al. 2013).

Next to mineral nutrients, AM fungi can also improve the water supply of the host. Passive water flux across membranes along an osmotic gradient takes place via major intrinsic proteins, the aquaporins, which can also contribute to transport of ammonium, metalloids, glycerol, signaling molecules, and dissolved gases (Bienert & Chaumont 2014, Li et al. 2014). Expression of aquaporins can be altered during AM interaction, also depending on the water and the potassium status of plants (Uehlein et al. 2007, El-Mesbahi et al. 2012, Bárzana et al. 2014). In different plant species, induced gene expression was found for plasma membrane intrinsic proteins (PIPs) and/or Nodulin 26-like intrinsic proteins (NIPs) in mycorrhizal roots (Uehlein et al. 2007, Alguacil et al. 2009, Gomez et al. 2009, Hogekamp et al. 2011, Giovannetti et al. 2012). Nodulin 26 is an LRS-specific aquaporin localized exclusively to the symbiosome membrane of rhizobia-infected cells, where it accounts for around 15% of the protein, facilitating the transport of ammonia, water, and glycerol (Wallace et al. 2006). Accordingly, laser microdissection revealed that *NIP1* genes of *L. japonicus* and *M. truncatula* are specifically expressed in arbusculated cells (Gomez et al. 2009, Hogekamp et al. 2011, Giovannetti et al. 2012), and immunolabeling indicates a *de novo* accumulation of (plasma membrane intrinsic) aquaporins in the root cortex of colonized plants (Uehlein et al. 2007). In heterologous expression systems, the AM-induced MtPIP2;1 and LjNIP1 clearly increased water permeability, whereas MtNIP1 was demonstrated to act as low affinity ammonium transporter (Uehlein et al. 2007, Giovannetti et al. 2012). Thus, the periarbuscular membrane is suggested to contain an elevated number of aquaporins mediating the exchange of water and ammonium and potentially also of signaling compounds like hydrogen peroxide.

The prebranching cytoplasmic aggregations that were found in the inner cortical PPA to precede hyphal branching are suggested to be of major importance for targeting the symbiosis-relevant host nutrient transporters to the periarbuscular membrane (Genre et al. 2008). They are also hypothesized to play a role in generating the putative membrane-derived lysophospholipid signal that elicit expression of AM-inducible phosphate transporter genes (Drissner et al. 2007, Genre et al. 2008).

3. Regulation of the AM Symbiosis by Carbon and by Plant Apoplastic Invertases

The AM fungi are obligate biotrophs that rely in their carbon supply on the living host. For the plant partner, this interaction can account for a major carbon-cost that is predicted to amount from 4 to 20 % of the photosynthetically fixed carbon (for overview see Douds et al. 2000, Graham 2000). Thus makes carbon to a major regulator in the AM symbiosis to limit the plant's cost, in dependence of its benefits.

3.1. Interconnection of Mineral and Carbon Exchange

Recent analyses of the mineral nutrient fluxes clearly indicate that the carbon flow from the plant to the fungus is interrelated with the delivery of mineral nutrients towards the plant – and vice versa (Hammer et al. 2011, Kiers et al. 2011, Fellbaum et al. 2012, 2014). Previous studies already showed that light shortage of plants can reduce AM fungal root colonization as well as phosphate inflow upon mycorrhization (Hayman 1974, Tester). Using a *D. carota* root culture co-cultivated with *R. irregularis* in a two-compartment system (comprising a mycorrhizal root compartment and a fungal compartment) and inductively coupled plasma atomic emission spectroscopy revealed that under low carbon conditions within the root compartment, phosphorus accumulates in the fungus (Hammer et al. 2011). In contrast, increased C availability of the host plant can stimulate the AM fungal uptake of P and N and their translocation to the host as verified with isotope-labeled compounds in such a two-compartment co-culture (Bücking & Shachar-Hill 2005, Fellbaum et al. 2012). Thus, carbon is an important trigger for fungal uptake and transport of mineral nutrients.

In ecosystems, however, plants are usually colonized by several AM fungi which in return interconnect a considerable number of plants forming belowground mycorrhizal networks and which provide more or less benefits with different C cost for the plant (van der Heijden & Horton 2009, Lendenmann et al. 2011). Here, not only carbon competition between one plant and one fungus occurs, but resource competitions between multiple fungi and between multiple plants. On the one hand, when plants were infected with multiple AM fungal species, carbon was preferentially allocated to the most cooperative species (Kiers et al. 2011). The authors found *Glomus aggregatum* to be less cooperative than *R. irregularis* retaining, upon sucrose application to the root compartment, significantly more P in form of long-chained polyphosphates that are unavailable for the host (Kiers et al. 2011). Accordingly, upon infection with both fungi, less carbon was allocated to *G. aggregatum* compared to *R. irregularis* and also upon triple infection the cooperative species received more carbon. If roots are only infected with *G. aggregatum*, no additional carbon was allocated to the fungus upon P application to the fungal compartment, different to the data achieved for *R. irregularis*-infected roots (Kiers et al. 2011) indicating the fine-tuned carbon delivery according to the benefit sensed by the plant. On the other hand, usually also multiple plants compete for resources supplied by the fungus. Using an experimental setup with different combinations of shaded and non-shaded *M. truncatula* plants demonstrated that also AM fungi are able to discriminate between plants with different source strengths (Fellbaum et al. 2014). *G. aggregatum* as well as the more cooperative *R. irregularis* preferentially allocated nutrient resources to non-shaded host plants that were able to provide more carbon to the fungal partner (Fellbaum et al. 2014). This indicates that both, plant host and AM fungal microsymbiont, can sense the benefits provided by different partners and that the interaction is controlled and stabilized in a bidirectional way by reciprocal rewarding. In plants with an ineffective symbiotic P uptake pathway, the AM fungal partner is unable to fully establish within the root as demonstrated by knock-down and/or mutation of the AM-specific phosphate transporter genes *LjPT3* and *MtPT4* (Maeda et al. 2006, Javot et al. 2007a). Here the interaction is blocked by induction of defense mechanisms and/or C starvation of the fungal partner (see also subsection 2.3.3.).

3.2. Carbon Cycling in the AM Interaction and the Role of Plant Sucrose-Degrading Enzymes

During the asymbiotic and presymbiotic phases, AM fungi feed their growth with carbon reserves stored in their large spores in form of triacylglycerol (Gaspar et al. 1994). In addition, germinating spores can take up and metabolize some C compounds as shown for acetate, glucose, fructose, and glycerol (but not mannitol) (Bago et al. 1999, Lammers et al. 2001, Bücking et al. 2008), which might contribute to facilitate multiple germination events discontinued by resting of the spores if no suitable plant partner is available. Interestingly, the uptake of exogenous (isotope-labeled) organic carbon by germinating spores could be stimulated by root exudates (Bücking et al. 2008). In addition, CO_2 is suggested to act as a carbon source during presymbiotic fungal growth (Bécard & Piché 1989). At the presymbiotic phase, AM fungi showed general ability to synthesize trehalose but they appear to lack any considerable synthesis of fatty acids and storage lipids (Bago et al. 1999). To fulfill their lifecycle, the obligate biotrophic AM fungi have to interact with a plant partner.

During the symbiotic phase, the AM fungi rely on plant-derived carbon. According to other plant-fungus interactions including interactions with ectomycorrhizal fungi and biotrophic pathogens (see e.g. Nehls & Hampp 2000, Mendgen & Hahn 2002, Nehls et al. 2010, Talbot 2010), sugars have been strong candidates for the form in which carbon is transferred from the host to the fungus. Using isotope-labeled compounds, substantial uptake of glucose via the AM fungal intraradical mycelium was demonstrated, whereas fructose uptake was considerably lower and mannitol and succinate were not found to be taken up (Shachar-Hill et al. 1995, Solaiman & Saito 1997, Pfeffer et al. 1999). The proposed role of glucose as plant-derived carbon and energy source to AM fungi was also supported by a relatively high hexokinase activity detected in intraradical hyphae (Saito 1995). After uptake, glucose becomes rapidly converted within the intraradical mycelium into trehalose and glycogen as first carbon pool, probably to buffer intracellular glucose levels (Shachar-Hill et al. 1995, Pfeffer et al. 1999). This is in line with an up-regulation of glycogen synthase genes during the symbiotic phase (Bago et al. 2003, Tisserant et al. 2013). The intraradical mycelium is also the site of lipid synthesis. Carbon is stored and transported within the fungus in form of triacylglycerol as well as glycogen, both are also exported to the extraradical mycelium (Bago et al. 2002, 2003) to support hyphal growth in the soil and formation of new spores that contain a large amount of triacylglycerol as carbon reserve for germination and presymbiotic growth (Bago et al. 2002). Different to intraradical structures, the extraradical mycelium appeared to be unable to take up hexoses (including glucose) (Pfeffer et al. 1999); however, an uptake of acetate and glycerol could be demonstrated (Lammers et al. 2001). A considerable amount of the carbon received from the host is released by the extraradical AM fungal structures to the soil (and partially to the atmosphere), particularly in form of the glycoprotein glomalin (Rillig 2004) and as CO_2 (Johnson et al. 2002).

The carbon supply of the AM fungus increases the sink strength of the root upon mycorrhization. In plants, sucrose is the major form for phloem-mediated long-distance carbon and energy transport from the photosynthetic active source leaves to the heterotrophic sink organs (Lemoine et al. 2013). The efflux and uptake of sugars across membranes can take place passively via a concentration gradient or by active transporter proteins, usually H^+ symporter or antiporter. Upon mycorrhization of *S. lycopersicum*, plant sucrose transporter genes were found to be up-regulated in leaves, probably to facilitate phloem loading and thus sucrose translocation to the mycorrhizal roots (Boldt et al. 2011, Doidy et al. 2012). Such sucrose transporter genes were found to be inducible upon low-P conditions (Doidy et al. 2012). An increased allocation of C resources to the root under P starvation might promote and support an AM fungal symbiosis by providing the required energy and also by sugar-mediated signaling (for overview see Hammond & White 2008, 2011, Lei & Liu 2011). In addition to sucrose transporters in leaves, plant sugar transporters can be also locally up-regulated in mycorrhizal roots, such as the *M. truncatula* hexose transporter Mtst1 (Harrison 1996) and sucrose transporters of *M. truncatula* and *S. lycopersicum* (García-Rodríguez et al. 2007, Boldt et al. 2011, Doidy et al. 2012, Gaude et al. 2012, Bitterlich et al. 2014). Those transporters are suggested to support the root cells with sufficient carbon. Such a competition with the endosymbiont for C compounds might explain the increased mycorrhizal colonization and enhanced formation of fungal storage organs in *S. lycopersicum* plants with suppressed root expression of *SUT2*, whose gene product is in wild-type roots localized at the periarbuscular membrane (Bitterlich et al. 2014). Moreover, SlSUT2 appears to be involved in brassinosteroid signaling or biosynthesis.

Because AM fungi appear to lack sucrose-cleaving enzymes (see also Tisserant et al. 2012, 2013), corresponding plant enzymes are suggested to have a crucial function in delivering C compounds in form of hexoses that can be taken up by the intraradical AM fungal structures. In plants, sucrose can be cleaved by sucrose synthases and invertases. Sucrose synthases, which are present in the cytosol, reversibly convert sucrose into fructose and UDP-glucose. Invertases cleave sucrose into glucose and fructose and can be classified in accordance to their optimal pH value and subcellular localization into three groups: (i) the soluble alkaline invertases associated with the cytoplasma, (ii) the soluble acid invertases located in the vacuole, and (iii) the acid apoplastic invertases that are ionically bound to the plant cell wall (for overview see Tymowska-Lalanne & Kreis 1998, Sturm & Tang 1999, Roitsch & González 2004, Fotopoulos 2005). Upon AM interaction, members of all types of plant sucrose-cleaving enzymes were found to exhibit induced transcriptional activation and/or elevated enzymatic activity in mycorrhizal roots tissue of different plant species (see Table 1).

Table 1. Induction of plant sucrose-cleaving enzymes upon AM interaction. (The table is continued on the next page.)

Enzyme type / Plant species	AM fungal species	Gene	Type of induction	Weeks after inoculation	Reference
Sucrose synthases:					
T. repens	natural mixture		E↑ in mycorrhizal roots	5 to 12	Wright et al. (1998)
G. max	*Funneliformis mosseae*		E↑ in mycorrhizal roots	~6	Schubert et al. (2003)
Z. mays	*Rhizophagus intraradices*, *F. mossea*	*Sus1*	T↑ in mycorrhizal roots	~1.5 to ~4	Ravnskov et al. (2003)
Z. mays	*R. intraradices*, *F. mossea*	*Sh1*	T↑ in mycorrhizal roots	~1.5 to ~2.5	Ravnskov et al. (2003)
P. vulgaris	*R. intraradices*	*Rhizobium*-responsive gene	T↑ in arbusculated cells	4	Blee & Anderson (2002)
S. lycopersicum	*R. intraradices*, *F. mossea*	*TOMSSF*	T↑ in mycorrhizal roots	4 and 6	García-Rodríguez et al. (2007)
S. lycopersicum	*Rhizophagus fasciculatus*	*Sus3*	T↑ in mycorrhizal roots	~6	Tejeda-Sartorius)
M. truncatula	*F. mossea*	*SucS1*	P↑ in arbusculated cells, T↑ in mycorrhizal root tissue	3.5	Hohnjec et al. (2003)
Cytosolic invertases:					
T. repens	natural mixture		E↑ in mycorrhizal roots	5 to 15	Wright et al. (1998)
G. max	*F. mosseae*		E↑ in mycorrhizal roots	~2 to 6	Schubert et al. (2003)
P. vulgaris	*R. intraradices*	*Pvsai*	T↑ in arbusculated cells	4	Blee & Anderson (2002)

Enzyme type / Plant species	AM fungal species	Gene	Type of induction	Weeks after inoculation	Reference
Vacuolar invertases:					
T. repens	natural mixture		E↑ in mycorrhizal roots	5 to ~9	Wright et al. (1998)
S. lycopersicum	*R. intraradices*, *F. mossea*	*TIV1*	T↑ in mycorrhizal roots	6	García-Rodríguez et al. (2007)
Apoplastic invertases:					
T. repens	natural mixture		E↑ in mycorrhizal roots	~9 to 15	Wright et al. (1998)
S. lycopersicum	*R. irregularis*		E↑ near arbuscules and intercellular hyphae	3-4	Schaarschmidt et al. (2006)
S. lycopersicum	*R. intraradices*, *F. mossea*	*LIN6*	T↑ in mycorrhizal roots	6	García-Rodríguez et al. (2007)
S. lycopersicum	*R. fasciculatus*	*LIN6*	T↑ in mycorrhizal roots	~6	Tejeda-Sartorius)
S. lycopersicum	*R. irregularis*	*LIN6*	P↑ and T↑ near arbuscules and intercellular hyphae	10, 11	Schaarschmidt et al. (2006)
Invertases (not further classified):					
P. vulgaris	*Claroideoglomus etunicatum*		E↑ in mycorrhizal roots	5	Dehne (1986)
P. vulgaris	*R. intraradices*		E↑ in mycorrhizal roots	4	Blee & Anderson (2002)
Citrus junos	*F. mosseae*		E↑ of acid invertases in mycorrhizal roots	~12	Li et al. (2012)

P: promoter activity; T: transcript accumulation; E: enzyme activity; ↑: up-regulation compared to non-mycorrhizal tissue.

The parallel analyses of the activity of individual enzyme types in mycorrhizal white clover (*Trifolium repens*) roots revealed differences in their induction over time. Here, activities of sucrose synthases, cytoplasmic invertases, and vacuolar invertases of mycorrhizal roots were higher compared to non-mycorrhizal plants at earlier stages with lower (or missing) enhancement at late time-points, whereas the cell wall invertase activity further increased over time exhibiting enhanced activities compared to non-mycorrhizal plants also at late stages (Wright et al. 1998). An early increase in the cytosolic invertase activity, starting already with the beginning of AM fungal root colonization, was also found in *G. max* roots (Schubert et al. 2003). This indicates a rapid mobilization of cellular carbon pools, probably to feed the intracellular rearrangements of the host cells, and an increasing sink strength of the colonized root over time, which might be forced by apoplastic invertase-mediated phloem unloading (Tymowska-Lalanne & Kreis 1998, Roitsch et al. 2003). Extracellular invertases are also likely candidates for providing apoplastic hexoses for AM fungal uptake from the symbiotic interface.

In mycorrhizal *S. lycopersicum* roots, an *in situ* activity staining indeed revealed, already few weeks after inoculation, locally increased invertase activities in the root phloem and around arbuscules and intercellular hyphae (Schaarschmidt et al. 2006). However, analyzing whole root extracts, no elevated invertase activities could be detected even at later stages (up to 10 weeks after inoculation). This corresponds to a dilution effect if changes occur only locally causing general detection problems when whole tissues or organs are analyzed. Increases in apoplastic invertase activity can be caused by acidification of the apoplastic compartment, which is estimated to be usually between 5 and 6.5 (Grignon & Sentenac 1991, Felle 1998, Yu et al. 2000, Gao et al. 2004), since these enzymes have an pH optimum between approximately 4 to 5 (Wright et al. 1998, Roitsch & González 2004). The low pH of the periarbuscular interface that is estimated to be around 4 (Guttenberger 2000) can contribute to the invertase activity detected near arbuscules (Schaarschmidt et al. 2006). In addition, *de novo* synthesis is likely to occur, including apoplastic invertases in the symbiotic interface. Transcriptional activation of an apoplastic invertase gene in mycorrhizal roots was first demonstrated by Schaarschmidt and co-workers) for the *S. lycopersicum* apoplastic invertase gene *LIN6*. LIN6 belongs to a gene family of four members and has a major role in regulating sink strength in actively growing sink tissues including seedling roots and in response to several stress-related external stimuli (Godt & Roitsch 1997, Goetz et al. 2000, Sinha et al. 2002,

Thoma et al. 2003). Corresponding to the predicted role in phloem unloading and AM fungal hexose supply, increased promoter activation and transcript accumulation upon mycorrhization was detected in the root vasculature and near intra- and intercellular fungal structures, which corresponds to the detected invertase activity (Schaarschmidt et al. 2006). An enhanced *LIN6* transcript accumulation in mycorrhizal roots was also shown in later studies (García-Rodríguez et al. 2007, Tejeda-Sartorius et al. 2008). In addition to the *S. lycopersicum LIN6* gene, elevated transcript levels in mycorrhizal roots were also found for one vacuolar and one cytoplasmic invertase gene of *P. vulgaris* (Blee & Anderson 2002) and *S. lycopersicum* (García-Rodríguez et al. 2007), respectively and for sucrose synthase genes of *M. truncatula*, *P. vulgaris*, maize (*Zea mays*), and *S. lycopersicum* (Blee & Anderson 2002, Hohnjec et al. 2003, Ravnskov et al. 2003, García-Rodríguez et al. 2007, Tejeda-Sartorius et al. 2008).

However, less was known about the actual role of AM-inducible plant sucrose-cleaving enzymes for the AM symbiosis. Comprehensive studies on the sucrose synthase gene *SucS1* from *M. truncatula* gave new insights into its function during AM symbiosis and LRS (Hohnjec et al. 2003, Baier et al. 2007, 2010). *MtSucS1* antisense knock down lines were characterized by impaired nodulation and mycorrhization with inefficient nutrient exchange and by reduced plant growth and development upon interaction with the endosymbionts (Baier et al. 2007, 2010). These phenotypes might be caused by deprivation in energy and carbon resources required for cellular rearrangements in the host. For invertases, locally reduced apoplastic invertase activity achieved by root-specific expression of an acid invertase inhibitor led as well to a reduced AM fungal colonization (Figure 5a) (Schaarschmidt et al. 2007a). This is in line with the predicted role of plant cell wall-bound invertases in hexose delivery to the apoplastic fungal structures. However, in contrast to that, root-specifically enhanced apoplastic invertase activity was not sufficient to stimulate root colonization, despite elevated root hexose levels (Figure 5b) (Schaarschmidt et al. 2007a). Thus, it is supposed that sucrose-cleaving enzymes, particularly apoplastic invertases, are usually sufficiently induced upon AM interaction. Moreover, their induction and activity are usually tightly regulated, including inhibition by products and by inhibitor proteins (Rausch & Greiner 2004), since apoplastic invertases are also known to be involved in activation of plant defense responses by providing hexoses as energy source and as signaling molecules (Roitsch et al. 2003, Proels & Hückelhoven 2014).

In the compatible interaction of pepper (*Capsicum annuum*) with the bacterial pathogen *Xanthomonas campestris*, gene expression of a plant apoplastic invertase gene is suppressed by microbial effectors that are released by a type three secretion system (T3SS) suppressing host immune responses (Sonnewald et al. 2012). Infection with a T3SS-deficient strain led to strong induction of apoplastic invertase followed by *PR* gene expression and caused no visible symptoms. Induction of plant apoplastic invertases was also found e.g. upon pathogenic infection of tobacco (*Nicotiana tabacum*) leaves with potato virus Y (PVY) (Herbers et al. 2000) or with the hemibiotrophic oomycete *Phytophthora nicotianae* (Essmann et al. 2008), of barley (*Hordeum vulgare*) leaves with the biotrophic fungus *Blumeria graminis* (Swarbrick et al. 2006), of *S. lycopersicum* leaves with the necrotrophic fungus *Botrytis cinerea* or with the biotrophic bacterium *Pseudomonas syringae* (Berger et al. 2004), and of *D. carota* roots with the necrotrophic bacterium *Erwinia carotovora* (Sturm & Chrispeels 1990). In all these cases, apoplastic invertase induction was accompanied or followed by transcript accumulation of pathogenesis-related (*PR*) or of phenylalanine ammonium lyase (*PAL*) gene(s) (for PVY see also Herbers et al. 1996).

Posttranslational inhibition of apoplastic invertase activity in leaves of *A. thaliana* was shown to increase the susceptibility to *P. syringae* (Bonfig et al. 2006). RNAi-mediated knock-down of apoplastic invertase in *N. tabacum* impaired defense mechanisms lowering *PR* gene expression, PAL activity, and hypersensitive cell death upon *P. nicotianae* infection (Essmann et al. 2008). Accordingly, overexpression can cause stimulation of plant defense mechanisms such as accumulation of *PR* gene transcripts, salicylic acid, and phenolic compounds and can lower the susceptibility to PVY (Herbers et al. 2000, Baumert et al. 2001). Here, overexpression was achieved by expression of a chimeric gene encoding an apoplast-located yeast-derived invertase under the CaMV 35S promoter (von Schaewen et al. 1990). In homozygous plants, this overexpression results in a strong accumulation of hexoses in the source leaves which in return down-regulate photosynthetic activity and reduce translocation of sugars to the sink organs – all resulting in stunted plant growth (von Schaewen et al. 1990, Sonnewald et al. 1991).

The effect of constitutively elevated apoplastic invertase activity on the AM symbiosis was analyzed in plants with heterozygous expression of the yeast-derived apoplast-located invertase (Schaarschmidt et al. 2007b). In plants characterized by a strong apoplastic invertase activity that particularly occurred in the shoot, AM fungal root colonization was reduced, probably due to sugar depletion of the sink roots and activation of defense mechanisms

(Figure 5a). Astonishing, in plants that were characterized by an only moderately enhanced apoplastic invertase activity in the shoot, mycorrhization was stimulated (Figure 5b). Such plants showed only marginal phenotypical effects and no elevated hexose accumulation in the shoot could be detected. Corresponding to that, *PR* gene expression was not induced; instead, the root had even lower contents of defense-related metabolites, such as chlorogenic acid and scopolin (Schaarschmidt et al. 2007b) that accumulate upon PVY infection and in homozygous apoplastic invertase expressing plants (Baumert et al. 2001). Moreover, the plants with moderate invertase induction had higher root levels of ABA (Schaarschmidt et al. 2007b) that was previously shown to positively affect AM formation (Herrera-Medina et al. 2007, Martín Rodriguez et al. 2010).

An endogenous (developmental independent) rise in shoot apoplastic invertase activity can be caused in response to different external abiotic and biotic factors and is often mediated by changes in the phytohormone homeostasis. For example, wounding of plant tissue led to a rise in jasmonate levels and both, wounding and application of methyl jasmonate (MeJA), were found to induce plant apoplastic invertases (Zhang et al. 1996, Godt & Roitsch 1997, Thoma et al. 2003, Schaarschmidt et al. 2006). Repeated but slight leaf wounding of *M. truncatula* was accompanied by a stimulation of JA biosynthesis and by an induction of few invertase genes, including one predicted apoplastic invertase that was found to be inducible by MeJA, and led to a reduced above- and belowground growth indicating the higher energy demand of wounded plants (Landgraf et al. 2012). Similar to *N. tabacum* plants with moderately elevated apoplastic invertase activity in the shoot, mycorrhization in the slightly leaf-wounded *M. truncatula* plants was stimulated and was able to partly counteract the negative effects of wounding. In contrast, such a wounding treatment reduced root colonization by the biotrophic oomycete *Aphanomyces euteiches* (Landgraf et al. 2012). The reduced pathogen infection might be caused by JA-induced defense mechanism and carbon competition, which, astonishing, did not negatively affect root colonization by the AM fungus *R. irregularis* (Landgraf et al. 2012). This indicates specific regulations of mycorrhizal and pathogenic plant–microbe interactions, probably by sensing the benefits provided by AM fungi.

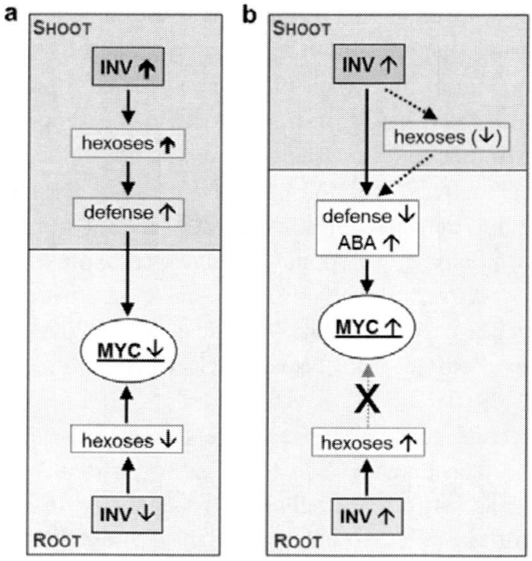

Figure 5. Regulation of AM fungal colonization by modulated apoplastic invertase activity in the root or in the shoot. a: Reduced colonization was found in plants with lowered apoplastic invertase activity in the root that was accompanied by reduced glucose and fructose levels in the root tissue (Schaarschmidt et al. 2007a). Root-specific reduction of apoplastic invertase activity was achieved by expression of an acid invertase inhibitor under the *pyk10* promoter. Also, plants with a strongly increased apoplastic invertase activity in the shoot showed a lower degree of mycorrhization (Schaarschmidt et al. 2007b). Here, the expression of a yeast-derived invertase that is translocated to the apoplast via specific signal peptides (von Schaewen et al. 1990) strongly increased the shoot apoplastic invertase activity resulting in an accumulation of glucose and fructose in the source leaves and a suffering of the sink tissue. In addition, the excessive hexose content resulted in an activation of defense mechanisms as indicated by transcript accumulation of *PR* genes in the source leaves (Schaarschmidt et al. 2007b). b: A stimulated AM interaction and higher root colonization was found in plants with only slightly increased shoot apoplastic invertase activity caused by expression of the apoplast-located yeast invertase (Schaarschmidt et al. 2007b). Here, hexose levels in the shoot were not elevated, but either not altered or even lowered. These plants were characterized by a down-regulation of defense mechanisms, including reduced levels of phenolic compounds, and an increased ABA content in roots (Schaarschmidt et al. 2007b). In contrast, a root-specifically enhanced apoplastic invertase activity, achieved by local induction of an alcohol-inducible promoter system (Schaarschmidt et al. 2004) controlling expression of the apoplast-located yeast-derived invertase, could not stimulate AM fungal root colonization, although roots exhibited elevated glucose and fructose levels (Schaarschmidt et al. 2007a). INV: apoplastic invertase activity; MYC: AM fungal root colonization.

3.3. Site of AM Fungal C Uptake and Identification of the First AM Fungal Monosaccharide Transporter

The general regulation of the AM symbiosis by carbon is also mirrored in the AM morphology including the extraradical mycelium (Olsson et al. 2014) and the location of the intraradical fungal structures within the root. To enable a plant-controlled release of C compounds to the endosymbiont, AM fungi are not allowed to pass the endodermis and cannot reach the central core. Instead, arbuscules and coiled hyphae, as main nutrient exchange organs, are formed predominantly in the inner root cortex cells and thus as close as possible to the root phloem, suggesting an optimal position for the uptake of sugars (Blee & Anderson 1998). Hence, these fungal structures are predicted to play also a key role in the uptake of C compounds within the root. This was further supported by a correlation between arbuscule and spore formation (Douds 1994). A functioning of arbuscules and coils in both, the uptake of C compounds and the delivery of minerals might facilitate a coordinated nutrient exchange between the partners and the reciprocal rewarding of host and microsymbiont. In addition to the intracellular AM fungal structures, also intercellular hyphae appear to some extent to be involved in carbon uptake. Transcript accumulation and elevated activity of apoplastic invertase was not only found around arbuscules, but also near intercellular hyphae (Schaarschmidt et al. 2006). A similar expression pattern was found for the AM-inducible hexose transporter Mtst1 of *M. truncatula* (Harrison 1996). This transporter is suggested to facilitate the uptake of apoplastic hexoses by cortex cells in close vicinity to AM fungal structures including intercellular hyphae indicating carbon competition between the host and the endosymbiont and thus also an hexose uptake by the intercellular hyphae. Moreover, the fungal plasma membrane of intercellular hyphae was found to exhibit ATPase activity (Gianinazzi-Pearson et al. 1991), which is an indicator for active transport processes.

In contrast to the AM-specific phosphate transporters, the AM fungal transporters involved in the fungal uptake of C compounds remained completely unknown for a long time. The first glomeromycotan monosaccharide transporter, MAST1, was identified in *Geosiphon pyriformis*, a member of the Glomeromycota that undergo a symbiotic interaction with the cyanobacterium *Nostoc punctiforme* (Schüßler et al. 2006). GpMAST1 functions as an H^+ co-transporter with highest affinity to glucose; some uptake was also found for mannose, galactose, fructose and xylose, but not for sucrose. Recently, the first monosaccharide of an AM fungus was identified and characterized (Helber et al. 2011). The monosaccharide gene *MST2* of *R. irregularis* is almost exclusively expressed in intraradical fungal structures and transcripts were found in arbuscules as well as intercellular hyphae upon interaction with a *S. tuberosum* hairy root culture. The encoded membrane-bound H^+ co-transporter shows high affinity to glucose with pH optimum at pH 5 or lower. AM fungal ATPase genes, including one that is almost exclusively expressed during the symbiotic phase, were previously identified (Requena et al. 2003). In the heterologous yeast system, uptake of glucose by RiMST2 could be efficiently outcompeted by plant cell wall components, namely mannose, galactose, glucuronic acid, galacturonic acid, and xylose (Helber et al. 2011). Xylose was previously found to be transported by some fungal monosaccharide transporters including a low rate transport by the glomeromycotan GpMAST1 (Hamacher et al. 2002, vanKuyk et al. 2004, Schüßler et al. 2006). The uptake capacity for such cell wall constituents is in agreement with the composition of the symbiotic containing non-crosslinked cell wall components (see also subsection 2.3.2.) and indicates alternative plant-derived carbon sources in addition to glucose.

Interestingly, *RiMST2* expression in the extraradical mycelium was induced by xylose (Helber et al. 2011). However, different to a previous study (Pfeffer et al. 1999), the extraradical mycelium was found to exhibit uptake capacity for glucose and xylose even without *RiMST2* induction by xylose. This uptake might occur by another predicted monosaccharide transporter, encoded by *RiMST4*, which was constitutively expressed in the extraradical mycelium (Helber et al. 2011). A third predicted monosaccharide transporter gene of *R. irregularis*, *RiMST3*, show only slight transcript accumulation in intraradical structures. The expression of *RiMST2* in the intraradical fungal structures was suppressed by phosphate in a similar manner than the AM-inducible *S. tubersoum PT4* gene, supporting the predicted interconnection of intraradical carbon and phosphate exchange. Using an elegant host-induced gene silencing (HIGS) approach, Helber and co-workers (2011) were able to demonstrate the requirement of *RiMST2* for successful AM symbiosis including full development of arbuscules without exhibiting premature senescence. Upon *RiMST2* silencing, *StPT4* expression was almost completely abolished.

Next to glucose and cell wall constituents, previous labeling experiments indicate glycerol as additional carbon source for AM fungi. Studies with ^{13}C-labeled glycerol demonstrated that in addition to germinating spores and extraradical mycelium of AM fungi (Lammers et al. 2001), also intraradical structures are able to take up and metabolize glycerol (Bago et al. 2002).

Interestingly, Nodulin 26-like aquaporins, NIPs, are known to be able to facilitate the transport of glycerol and a few other small molecules across membranes (see also subsection 2.3.3.). Nodulin 26, an aquaporin of *G. max* exclusively found on the symbiosome membrane during LRS and one of the first aquaglyceroporins documented in plants, has the ability to flux ammonia, formamide, and uncharged polyols such as glycerol (Rivers et al. 1997, Dean et al. 1998, Wallace et al. 2006). Interestingly, it shows distinct functional selectivity for glycerol permeability and only a weak water transport rate which is substantially lower compared to the archetype of water-selective aquaporins, the human AQP1 (Rivers et al. 1997, Wallace et al. 2002, Guenther et al. 2003). In the LRS, next to ammonia transport, Nodulin 26 is supposed to be involved in osmoprotection of the symbiosomes that are highly sensitive to osmotic stress (for overview see Wallace et al. 2006). In the AM interaction, different NIPs and plasma membrane intrinsic PIPs were found to be induced upon root colonization and to be localized to the periarbuscular membrane, probably to support the water and nutrient uptake by the plant cell (Uehlein et al. 2007, Gomez et al. 2009, Hogekamp et al. 2011, Giovannetti et al. 2012). Glycerol permease activity is less known for PIPs, but was demonstrated for a *Chlamydomonas reinhardtii* PIP (Anderca et al. 2004) and interestingly also for the AM-induced MtPIP2;1 (Uehlein et al. 2007). For NIPs, glycerol transport appears as a general feature (Weig & Jakob 2000, Wallace et al. 2002, Cabello-Hurtado & Ramos 2004). NIPs are suggested to be acquired via horizontal gene transfer from bacteria at the origin of plants (Zardoya et al. 2002) and were probably recruited to transport glycerol in plants since absence of (other) aquaglyceroporins. The solute flux across Nodulin 26 appears to occur by a pathway that is distinct from that for water (Rivers et al. 1997) and similar results were found for other aquaporins indicating that (bidirectional) water and solute transfer might also occur in opposite directions (Meinild et al. 1998, Bienert et al. 2008).

Thus, it is tempting to speculate that in addition to an uptake of water and ammonia by the plant cell, the AM-induced NIPs (and PIPs) also facilitate a glycerol flux from the plant towards the AM fungus, similar to that predicted to occur in the LRS via Nodulin 26. AM fungi were previously shown to be able to take up and metabolize glycerol (Pfeffer et al. 1999). It would be interesting to test, whether the identified monosaccharide transporter *RiMST2* is also capable in taking up glycerol. The uptake of carbon compounds other than glucose (and fructose) might also explain why elevated apoplastic invertase activity in the root could not stimulate AM fungal root colonization despite elevated glucose and fructose levels in the root (Schaarschmidt et al. 2007a).

4. Systemic Feedback Regulation of the AM Symbiosis by Autoregulation

As outlined above, the plant tightly controls its interaction with the AM fungal symbionts to reduce carbon losses and to maintain the mutualistic character of the symbiosis. The same is true for the interaction of legumes with nitrogen-fixing rhizobia that are also heterotrophic organisms. One common regulatory mechanism that limits the number of successful root infections is called autoregulation. This mechanism describes a feedback regulation that is somehow initiated by the common early signaling pathway required for infections with AM fungi and rhizobia (see subsection 2.2.).

The autoregulation was first identified for the LRS, called autoregulation of nodulation (AON) (for review see also Ferguson et al. 2010). Here, the effect of autoregulation is particularly apparent since every successful infection event results in the formation of a root nodule. A feedback inhibition of nodulation was already described in the early 1950ies demonstrating that nodule formation is controlled by the number of mature nodules and root tips (Nutman 1952). Mutant plants with defective autoregulation were first described for *G. max* and were identified by a super-nodulating phenotype with a strongly elevated number of root nodules over the entire root length and by a nitrate-tolerant nodulation (Carroll et al. 1985). Later, such mutants were found to be also characterized by a defective autoregulation of mycorrhization (AOM) that becomes visible by an enhanced AM fungal root colonization with higher arbuscule abundance as described for AON mutants of *G. max* (Shrihari et al. 2000, Sakamoto & Nohara 2009, Schaarschmidt et al. 2013), *L. japonicus* (Solaiman et al. 2000), *M. truncatula* and *P. sativum* (Morandi et al. 2000). These mutants were identified to have a defected CLAVATA1 (CLV1)-like receptor kinase, that is encoded by *NARK (Nodule Autoregulation Receptor Kinase)* in *G. max* (Searle et al. 2003), *HAR1* in *L. japonicus* (Krusell et al. 2002, Nishimura et al. 2002), *SUNN* (formerly named *Sym12*) in *M. truncatula* (Schnabel et al. 2005), and *SYM29* in *P. sativum* (Krusell et al. 2002). Grafting between mutant and wild-type plants demonstrated that this CLV1-like kinase, hereafter referred to as NARK, acts in the shoot of the plant during AON (Delves et al. 1986) and AOM (Sakamoto & Nohara 2009), revealing the existence of a long-distance signaling pathway in autoregulation (see also Figure 6). The corresponding systemic limitation of root infections in the entire root system was additionally proven by split-root experiments (Caetano-Anolles & Gresshoff 1990, Meixner et al. 2005, Kosslak & Bohlool 1984).

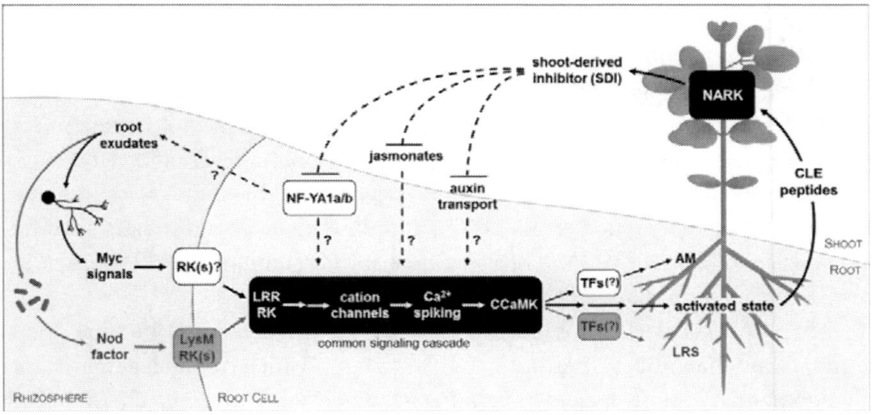

Figure 6. Model of the autoregulation system in legumes. The autoregulation of mycorrhization (AOM) and of nodulation (AON) are suggested to be initiated via the common early signaling cascade. This signaling pathway is required for successful AM fungal and rhizobial infections and in addition elicits a so-called activated state in the root resulting in the production of root-derived signal(s). For the AON, these are predicted to be CLE peptides (Reid et al. 2011). Corresponding signal(s) are also suggested for the AOM, but have so far not been identified. The root signal(s) are translocated to the shoot via the xylem. In the shoot, they activate a CLV1-like receptor kinase called NARK in *G. max* (Searle et al. 2003). Activation of NARK leads to the production of a shoot-derived inhibitor (SDI). SDI is predicted to mediate the NARK-dependent down-regulation of JA biosynthesis in the shoot (Kinkema & Gresshoff 2008), the reduced shoot-to-root auxin transport (van Noorden et al. 2006), and the transcriptional suppression of the homologous transcription factors NF-YA1a/b (Schaarschmidt et al. 2013). These NARK-dependent down-regulations are suggested to somehow negatively regulate element(s) the common early signaling pathway and/or production and release of specific root signals that can attract the microsymbionts. Thus finally results in a systemic suppression of subsequent infections by both endosymbionts.

The split-root system is a helpful tool to also uncover the effect of AOM in wild-type plants including the monocot *H. vulgare* (Pearson et al. 1993, Vierheilig et al. 2000b, Catford et al. 2003). In the AM symbiosis, the suppression of subsequent infections is less obvious compared to the LRS since the AM fungi are able to spread within the root without new epidermal infections. This ability can also mask the super-mycorrhization phenotype of *nark* mutants (see also Meixner et al. 2005, Schaarschmidt et al. 2013). Using split-root plants, the effect of a previous AM fungal inoculation of one root part on the subsequent infections of the other root part can be elucidated by the comparison to plants that lack a previous inoculation. Moreover, using split-root plants, the interconnection of AON and AOM was demonstrated (Catford et al. 2003). Using *Medicago sativa* it was shown that nodulation of one root part can suppress subsequent AM fungal colonization of the other root part and *vice versa*. Interestingly, Nod factor application was sufficient to trigger AON and AOM (Catford et al. 2003). This clearly indicates that the autoregulation is initiated by components of the common early signaling cascade. However, since the mycorrhization and nodulation phenotypes of some *nark* mutants appear to differ (Meixner et al. 2007), the existence of additional, specific elements in AOM and AON, correspondent to the commons and specifics in the establishment of both endosymbioses, were suggested.

The NARK-based autoregulation system is predicted to function in a similar way than the *A. thaliana* CLV system, which regulates meristem development by short-distance signaling (for overview see e.g. Clark 2001, Wang & Fiers 2010). In brief, in the *A. thaliana* CLV system CLV1 and CLV2 form a heteromer that biochemically interacts with the ligand CLV3. Ligand binding drives CLV1 phosphorylation, which promotes interaction with other molecules including a kinase-associated protein phosphatase (KAPP) that is a negative regulator of CLV1. The CLV system negatively controls the stem cell-promoting transcription factor WUSCHEL and by this regulates cell proliferation and differentiation in the shoot apical meristem. Corresponding to the *A. thaliana* system, potential KAPP (kinase-associated protein phosphatase) orthologs were identified in *G. max* showing capability to dephosphorylate NARK *in vitro* (Miyahara et al. 2008). An AtCLV2 ortholog was found to be encoded by *Sym28* in *P. sativum* (Krusell et al. 2011) and *sym28* mutants are characterized by super-nodulation (Sagan & Duc 1996) and super-mycorrhization (Morandi et al. 2000). Also PsSym28 is acting in the shoot (Sagan & Duc 1996) and is suggested to interact with the GmNARK ortholog PsSym29 during autoregulation. In *L. japonicus*, gene expression analyses of *CLV2* and *HAR1* (=*GmNARK* ortholog) indicate an overlapping organ expression patterns; however, a CLV2–HAR1 interaction could yet not been proven (Krusell et al. 2011). Moreover, a WUSCHEL-related homeobox transcription factor of *M. tuncatula* and *P. sativum*, WOX5, that is transcriptionally activated during nodule organogenesis appears to be controlled by NARK (Osipova et al. 2012). This indicates an involvement of the WOX–CLV system also in the control of cell proliferation and differentiation in the root nodule meristem; however, a function in the AM symbiosis has not been predicted (Osipova et al. 2012). An additional shoot component of the autoregulation system, at least for the AON, is encoded by *KLAVIER*, a predicted LRR receptor-like kinase identified in *L. japonicus* (Oka-Kira et al. 2005, Miyazawa et al. 2010). Recently, an involvement of the ubiquitin-dependent protein degradation pathway was suggested for the AON by identification of an ubiquitin fusion degradation protein-encoding gene, *UFD1a*, from *G. max*, which was in leaves found to be induced by rhizobia infection in a NARK-dependent manner (Reid et al. 2012).

Upon autoregulation, NARK is activated by root-derived signal(s) that are transported to the shoot most likely in the xylem sap (Djordjevic et al. 2007). In accordance to the *A. thaliana* CLV system, peptides similar to AtCLV3 are predicted as root-derived, NARK activating signals. In the last years, several CLV3/ESR-related (CLE) peptides were identified in legumes including nodulation- and nitrate-inducible ones (Oelkers et al. 2008, 2009, Okamoto et al. 2009, Mortier et al. 2010, Lim et al. 2011, Reid et al. 2011). Root-specific overexpression of such CLE peptides can suppress nodulation in a NARK-dependent manner (Okamoto et al. 2009, Lim et al. 2011, Reid et al. 2011, Osipova et al. 2012). The rhizobia-induced CLEs are supposed to act as root-derived signals in the AON activating NARK in the shoot, whereas nitrate-induced CLE peptides require action of NARK in the root (Reid et al. 2011). CLE peptides are also suggested to activate NARK in the AOM; however, AM-induced CLE peptides have yet not been described. NARK might be also activated by other or additional peptides. In *M. truncatula*, two peptide-encoding genes, *RALFL1* and *DVL1*, were identified by gene expression analysis of a super-nodulating double mutant after Nod factor treatment (Combier et al. 2008b). *MtRALFL1* and *MtDVL1* are induced upon Nod factor response and their overexpression resulted in reduced nodule number and enhanced abortion of infection threads. Thus makes MtRALFL1 and MtDVL1 likely candidates for root-derived signals involved in autoregulation. Activation of NARK in the shoot results in the production of a shoot-derived inhibitor (SDI). For the AON, SDI has been biochemically characterized as a heat-stable, ethanol-soluble, low-molecular weight molecule which is unlikely a RNA or protein (Kenjo et al. 2010, Lin et al. 2010). A putative receptor of SDI in the root might be encoded by *TOO MUCH LOVE* (*TML*) that was identified in *L. japonicus* by analyzing a novel super-nodulating mutant (Magori et al. 2009). Perception of the SDI somehow results in suppression of subsequent AM fungal and/or rhizobial infections (see also Figure 6).

Several studies point to an involvement of phytohormones in the AON, including auxins (van Noorden et al. 2006, 2007), cytokinins (Caba et al. 2000, Wopereis et al. 2000), and JA (Seo et al. 2007, Kinkema & Gresshoff 2008). Also for the AOM, an involvement of auxin is suggested since the local AM-induced increase in the root indole-3-acetic acid (IAA) content was lower in *G. max nark* mutants compared to the wild type (Meixner et al. 2005, 2007). However, in non-inoculated *M. truncatula sunn* mutant plants, a higher shoot-to-root auxin transport was detected in comparison to the corresponding wild type (van Noorden et al. 2006, 2007). Concerning JA, *nark* mutation was found to stimulate JA biosynthesis in the shoot compared to wild-type plants (Seo et al. 2007, Kinkema & Gresshoff 2008). (Moderately) elevated shoot JA levels are suggested to stimulate nodulation (Seo et al. 2007, Kinkema & Gresshoff 2008) and, as recently indicated by leaf-wounded *M. truncatula* plants, also the AM interaction (Landgraf et al. 2012). However, elevated JA biosynthesis in *nark* mutants appears to be development-dependent (Seo et al. 2007, Kinkema & Gresshoff 2008). Moreover, some application experiments with MeJA gave contradictory results indicating reduced calcium spiking and/or nodule number upon elevated jasmonate levels, even in *nark/har1* and *tml* mutants (Nakagawa & Kawaguchi 2006, Sun, Seo et al. 2007, Magori & Kawaguchi 2010). Such repressory effects might be (partly) explained by non-physiological jasmonate concentrations. In contrast, shoot-application of a JA biosynthesis inhibitor reduced super-nodulation in *nark* mutant plants, but not in the wild-type (Kinkema & Gresshoff 2008), which might support a role of JA during autoregulation (for review see also Hause & Schaarschmidt 2009). For cytokinins, Caba and co-workers (2000) found higher basal levels in roots of a *G. max nark* mutant compared to the wild type. However, studies on the cytokinin-insensitive *cre1* mutant of *M. truncatula* do not support a role for cytokinins in the AM interaction (see Foo et al. 2013 for this information); instead, cytokinins regulate, in interaction with ethylene, nodule organogenesis (Plet et al. 2011) and can also alter root growth which is affected in the *L. japonicus har1* mutant (Wopereis et al. 2000).

The systemic feedback inhibition of the AM interaction, that also occurs in non-leguminous plants (Vierheilig et al. 2000b, 2000c), appears to require a certain threshold of AM fungal colonization and/or a certain time-span to be activated (Vierheilig 2004). Different to the nitrate-tolerant nodulation phenotype of *nark* mutants (Carroll et al. 1985), the AOM was not found to be associated with the suppression of AM formation by phosphate (Vierheilig et al. 2000c, Schaarschmidt et al. 2013). Also C_{13} cyclohexenone apocarotenoids that accumulate upon AM fungal root colonization (Vierheilig et al. 2000a) were not found to be involved in the AOM in *H. vulgare* (Vierheilig et al. 2000c). Interestingly, autoregulation was accompanied by changes in the flavonoid pattern (Catford et al. 2006). Host plants can exude specific flavonoids to attract AM fungi and rhizobia (see subsection 2.1.1.). Some flavonoids were also found to inhibit presymbiotic AM fungal growth. In *M. sativa*, root part-specific inoculation with an AM fungus or with a rhizobium or Nod factor application resulted in changes of root exuded flavonoids (Catford et al. 2006). In all cases, a reduction of formononetin and ononin was found in the un-treated root part. For the LRS it was shown that application of formononetin or ononin to the subsequently inoculated root part could partly restore nodulation (Catford et al. 2006). Moreover, by applying root exudates, which were collected from different root systems, to cucumber (*Cucumis sativus*) plants inoculated with an AM fungus it could be demonstrated that only root exudates of non-mycorrhizal *C. sativus* plants, but not of mycorrhizal *C. sativus* plants (neither from the mycorrhizal nor from the non-mycorhrizal root-part) stimulate AM formation (Vierheilig et al. 2003). Root exudates of non-host plants either reduced AM fungal root colonization in *C. sativus* (as found for exudates of Brassicacae species) or had no effect (in case of *L. albus* root exudates) (Vierheilig et al. 2003). Upon AM fungal colonization, not only the pattern of root-exuded flavonoids changes, but also the exudation of strigolactones is reduced as indicated by the seed germination activity of parasitic weeds (Lendzemo et al. 2009, Louarn et al. 2012).

Such changes in the root exudates not only influence subsequent infections with AM fungi, but can also affected interactions with other organisms including parasites, indicating the role of AM fungi as potential biocontrol agent. However, different effects were found in different system. For example, root exudates from mycorrhizal strawberry and tomato plants reduced the sporulation of *Phytophthora fragariae* (Norman & Hooker 2000) and the attraction of *Phytophthora parasitica* zoospores (Lioussanne et al. 2009), respectively. *G. max* root exudates of mycorrhizal and/or nodulated plants suppressed growth of the soybean red crown rot pathogen *Cylindrocladium parasiticum* (Gao et al. 2012). Disease severity was further suppressed by activated defense mechanisms in mycorrhizal and/or nodulated roots indicated by accumulation of *PR* gene transcripts. In contrast, microconidia germination of the tomato pathogen *Fusarium oxysporum* f. sp. *lycopersici* was stimulated by root exudates of mycorrhizal *S. lycopersicum* plants (Scheffknecht et al. 2006). Interestingly, also plant growth-promoting rhizobacteria were more attracted by root exudates of mycorrhizal *S. lycopersicum* than of non-mycorrhizal plants (Gupta Sood 2003). Here, however, the effect was predicted to be caused by exuded organic acids or sugars, depending on the rhizobacterial species. In addition, the autoregulation can also affect infections with root-knot nematodes, since they are predicted to have recruited elements of the host response to mutualistic endosymbionts, particularly of the LRS, to elicit formation of giant cells in the host root (Weerasinghe et al. 2005). *L. japonicus har1* mutants were found to be hyper-infected by the root-knot nematode *Meloidogyne incognita* (Lohar & Bird 2003). Thus, in wild-type plants infected with rhizobia and/or AM fungi root-knot nematode infections might be suppressed via autoregulation. Further studies are required to elucidate commons and specifics in the AON and AOM as well as putative cross-talks with parasitic interactions.

To provide insights into the NARK-dependent transcriptome during AOM, recently the *G. max* gene expression of wild-type and *nark* mutants was analyzed in different tissues of split-root plants that either were partially inoculated with *R. irregularis* or stayed un-inoculated (Schaarschmidt et al. 2013). In previous studies, *G. max nark* mutant and wild-type plants were investigated during AON identifying the transcriptional regulation of JA-responsive and JA biosynthesis genes by NARK during the LRS (Kinkema & Gresshoff 2008) and the regulation of putative defense-related genes and predicted gibberellic acid biosynthesis genes (Hayashi et al. 2012). However, during AOM no obvious overlap was found with the gene expression during AON (Schaarschmidt et al. 2013). This might be due to different developmental stages of the plants or due to a specific signaling down-stream of NARK upon AON and upon AOM. During the AM interaction, only minor systemic changes were found compared to the local transcriptional reprogramming in mycorrhizal roots (Schaarschmidt et al. 2013). Among the genes verified to be regulated in a NARK-dependent manner during AM symbiosis was a predicted ornithine acetyl transferase gene, the stress-induced receptor-like kinase gene *GmSIK1,* and a predicted DEAD box RNA helicase gene (for more information see Schaarschmidt et al. 2013) – all reflected a systemic down-regulation of gene expression by NARK in roots and shoots. A predicted annexin gene, named *GmAnn1a*, was locally up-regulated by AM (independently of NARK) but was in the shoot suppressed by NARK (independently of AM). The expression pattern in different tissues indicates that shoot-specific suppression might particularly occur in the shoot vasculature (Schaarschmidt et al. 2013). This novel NARK-dependent expression pattern of an annexin gene indicates a role in long-distance signaling, next to the predicted function of symbioses-induced annexins in early intracellular signaling or in formation of the perifungal/symbiosome membrane (de Carvalho-Niebel et al. 2002, Manthey et al. 2004, Amiour et al. 2006, Talukdar et al. 2009, Schaarschmidt et al. 2013).

In addition, two predicted CCAAT-binding transcription factor subunit genes, named *GmNF-YA1a* and *GmNF-YA1b*, were systemically down-regulated by NARK in the root parts (Schaarschmidt et al. 2013). RNAi-mediated gene suppression of *GmNF-YA1a/b* in roots demonstrated a crucial role for these homologous transcription factors in AM establishment. Different to previously reported rhizobia-inducible *NF-Y* genes that might be targeted by NIN (see for example Combier et al. 2006, 2008a, Asamizu et al. 2008, Libault et al. 2009, Soyano et al. 2013), the repression of *GmNF-YA1a/b* gene expression upon AM interaction indicates a novel function of these genes in AON. It might be possible, that GmNF-YA1a/b target an element of the early signaling pathway or that they are involved in production of specific root-exuded signals (see also Figure 6). The specific role and targets of GmNF-YA1a/b and whether these transcription factors are also key players in the AON have to be further elucidated.

Next to gene expression changes upon autoregulation, also post-transcriptional regulation can occur to suppress or alter the endosymbiont-elicited transcriptional reprograming of the host that would be required for a successful root infection.

A transcriptome-wide degradome analysis performed on *M. truncatula* roots identified targets of miRNAs that undergo mRNA cleavage upon AM interaction (Devers et al. 2011). Among these, *NSP2* was identified whose transcripts were found to be cleaved upon interaction with a miRNA of the miR171 family, called miR171h (Devers et al. 2011). MiR171h expression is induced in response to Myc-LCOs and upon mycorrhization in the non-infected elongation zone of the root, which might be the target tissue for subsequent infections (Lauressergues et al. 2012).

Also CCAAT-binding transcription factors can be targets of miRNA-mediated gene silencing. In *M. truncatula*, the NF-YA gene *HAP2-1* that is involved in root nodule formation is negatively regulated by miRNA169 (Combier et al. 2006). And also in mycorrhizal roots, miR169d-triggered transcript cleavage was detected for one CCAAT-binding transcription factor gene (Devers et al. 2011). Weather such post-transcriptional regulations are dependent on NARK and are thus part of the autoregulation system has yet, however, not been investigated.

5. Looking Ahead

Since the first descriptions of mycorrhizal interactions starting in the 1840s including the innovative publications of Franciszek Kamieński and Bernhard Frank in the 1880ies (Kamienski 1881, Frank 1885), numerous studies has been performed on mycorrhizal symbioses, including the AM symbiosis, allowing fundamental insights. However, further research is required to fully understand the complex mechanisms that underlie the interaction between plants and AM fungi, which forms the most successful and ancient plant–microbe symbiosis. Most terrestrial plants still rely on this symbiosis to improve their fitness, although the AM symbiosis represents a considerable carbon-cost for plants which is guessed to account for up to 20% of the photosynthetically fixed carbon and an estimate of world-wide 5 billion tons of carbon annually (Bago et al. 2000). In return, the AM fungi contribute to improved (agro-)ecosystem productivity and diversity as well as soil quality and structure by accounting for most of the organic carbon including the secreted glue-like glycoprotein glomalin (Rillig 2004).

Establishing and maintaining a mutualistic AM symbiosis involves complex pathways and different mechanisms which are still not fully understood but which will provide important insights into the functioning of an AM symbiosis. Knowledge of this ancient and still widely-distributed plant–microbe interaction will also help to better understand the evolution and function of other plant–microbe interactions, including symbiotic and pathogenic ones that can as well have important economic and ecologic impacts. The recent sequencing of the transcriptome and the genome of the AM model fungus *R. irregularis* DAOM 197198 (Tisserant et al. 2012, 2013) is one key step in this direction. The sequence information will for example force the elucidation of the biotrophic character of the members of the Glomeromycota including their carbon uptake systems and the further identification of AM fungal signals by identifying the involved biosynthesis genes and analyzing their transcriptional regulation. Knowledge on the fungal genes and the according biosynthesis pathways can also help to differentiate between fungus- and host-derived origin of compounds detected in mycorrhizal roots.

Next to the research on model organisms including members of the Fabacea family and only few common AM fungal species, research on less characterized plant and fungal species will strongly contribute in further increasing our knowledge on AM functioning and putative conservation of mechanisms across the plant kingdom in both *Arum* and *Paris* type mycorrhizas. It will also help to elucidate the fungus- and plant-derived signals defining the AM morphology.

ACKNOWLEDGMENT

We acknowledge the excellent research performed on the arbuscular mycorrhizal symbiosis and which we did not cite here.

REFERENCES

Abdel-Lateif, K., Bogusz, D. & Hocher, V. (2012) The role of flavonoids in the establishment of plant roots endosymbioses with arbuscular mycorrhiza fungi, rhizobia and *Frankia* bacteria. *Plant Signal. Behav.* 7: 636-641.

Akiyama, K., Matsuoka, H. & Hayashi, H. (2002) Isolation and identification of a phosphate deficiency-induced C-glycosylflavonoid that stimulates arbuscular mycorrhiza formation in melon roots. *Mol. Plant Microbe Interact.* 15: 334-340.

Akiyama, K., Matsuzaki, K. & Hayashi, H. (2005) Plant sesquiterpenes induce hyphal branching in arbuscular mycorrhizal fungi. *Nature* 435: 824-827.

Akiyama, K., Ogasawara, S., Ito, S. & Hayashi, H. (2010a) Structural requirements of strigolactones for hyphal branching in AM fungi. *Plant Cell Physiol.* 51: 1104-1117.

Akiyama, K., Tanigawa, F., Kashihara, T. & Hayashi, H. (2010b) Lupin pyranoisoflavones inhibiting hyphal development in arbuscular mycorrhizal fungi. *Phytochemistry* 71: 1865-1871.Alexander, T., Meier, R., Toth, R. & Weber, H.C. (1988) Dynamics of arbuscule development and degeneration in mycorrhizas of *Triticum aestivum* L. and *Avena sativa* L. with reference to *Zea mays* L. *New Phytol.* 110: 363-370.

Alguacil, M.d., Kohler, J., Caravaca, F. & Roldán, A. (2009) Differential effects of *Pseudomonas mendocina* and *Glomus intraradices* on lettuce plants physiological response and aquaporin PIP2 gene expression under elevated atmospheric CO_2 and drought. *Microb. Ecol.* 58: 942-951.

Anderca, M.I., Suga, S., Furuichi, T., Shimogawara, K., Maeshima, M. & Muto, S. (2004) Functional identification of the glycerol transport activity of *Chlamydomonas reinhardtii* CrMIP1. *Plant Cell Physiol.* 45: 1313-1319.

Ane, J.-M., Levy, J., Thoquet, P., Kulikova, O., de Billy, F., Penmetsa, V., Kim, D.-J., Debelle, F., Rosenberg, C., Cook, D.R., Bisseling, T., Huguet, T. & Denarie, J. (2002) Genetic and cytogenetic mapping of DMI1, DMI2, and DMI3 genes of *Medicago truncatula* involved in Nod factor transduction, nodulation, and mycorrhization. *Mol. Plant Microbe Interact.* 15: 1108-1118.

Amiour, N., Recorbet, G., Robert, F., Gianinazzi, S. & Dumas-Gaudot, E. (2006) Mutations in *DMI3* and *SUNN* modify the appressorium-responsive root proteome in arbuscular mycorrhiza. *Mol. Plant Microbe Interact.* 19: 988-997.

Arrighi, J.-F., Barre, A., Ben Amor, B., Bersoult, A., Soriano, L.C., Mirabella, R., de Carvalho-Niebel, F., Journet, E.-P., Ghérardi, M., Huguet, T., Geurts, R., Dénarié, J., Rougé, P. & Gough, C. (2006) The *Medicago truncatula* lysin motif-receptor-like kinase gene family includes NFP and new nodule-expressed genes. *Plant Physiol.* 142: 265-279.

Asamizu, E., Shimoda, Y., Kouchi, H., Tabata, S. & Sato, S. (2008) A positive regulatory role for *LjERF1* in the nodulation process is revealed by systematic analysis of nodule-associated transcription factors of *Lotus japonicus*. *Plant Physiol.* 147: 2030-2040.

Bago, B., Pfeffer, P.E. & Shachar-Hill, Y. (2000) Carbon metabolism and transport in arbuscular mycorrhiza. *Plant Physiol.* 124: 949-958.

Bago, B., Pfeffer, P.E., Abubaker, J., Jun, J., Allen, J.W., Brouillette, J., Douds, D.D., Lammers, P.J. & Shachar-Hill, Y. (2003) Carbon export from arbuscular mycorrhizal roots involves the translocation of carbohydrate as well as lipid. *Plant Physiol.* 131: 1496-1507.

Bago, B., Pfeffer, P.E., Douds, D.D., Brouillette, J., Bécard, G. & Shachar-Hill, Y. (1999) Carbon metabolism in spores of the arbuscular mycorrhizal fungus *Glomus intraradices* as revealed by nuclear magnetic resonance spectroscopy. *Plant Physiol.* 121: 263-272.

Bago, B., Zipfel, W., Williams, R.M., Jun, J., Arreola, R., Lammers, P.J., Pfeffer, P.E. & Shachar-Hill, Y. (2002) Translocation and utilization of fungal storage lipid in the arbuscular mycorrhizal symbiosis. *Plant Physiol.* 128: 108-124.

Baier, M.C., Barsch, A., Küster, H. & Hohnjec, N. (2007) Antisense repression of the *Medicago truncatula* nodule-enhanced sucrose synthase leads to a handicapped nitrogen fixation mirrored by specific alterations in the symbiotic transcriptome and metabolome. *Plant Physiol.* 145: 1600-1618.

Baier, M.C., Keck, M., Godde, V., Niehaus, K., Küster, H. & Hohnjec, N. (2010) Knockdown of the symbiotic sucrose synthase MtSucS1 affects arbuscule maturation and maintenance in mycorrhizal roots of *Medicago truncatula*. *Plant Physiol.* 152: 1000-1014.

Balestrini, R. & Bonfante, P. (2014) Cell wall remodeling in mycorrhizal symbiosis: a way towards biotrophism. *Front. Plant Sci.* 5: 237.

Balestrini, R., Cosgrove, D.J. & Bonfante, P. (2005) Differential location of α-expansin proteins during the accommodation of root cells to an arbuscular mycorrhizal fungus. *Planta* 220: 889-899.

Balestrini, R., Gómez-Ariza, J., Lanfranco, L. & Bonfante, P. (2007) Laser microdissection reveals that transcripts for five plant and one fungal phosphate transporter genes are contemporaneously present in arbusculated cells. *Mol. Plant Microbe Interact.* 20: 1055-1062.

Balestrini, R., Hahn, M.G., Faccio, A., Mendgen, K. & Bonfante, P. (1996) Differential localization of carbohydrate epitopes in plant cell walls in the presence and absence of arbuscular mycorrhizal fungi. *Plant Physiol.* 111: 203-213.

Balzergue, C., Puech-Pagès, V., Bécard, G. & Rochange, S.F. (2011) The regulation of arbuscular mycorrhizal symbiosis by phosphate in pea involves early and systemic signalling events. *J. Exp Bot.* 62: 1049-1060.

Bárzana, G., Aroca, R., Bienert, G.P., Chaumont, F. & Ruiz-Lozano, J.M. (2014) New insights into the regulation of aquaporins by the arbuscular mycorrhizal symbiosis in maize plants under drought stress and possible implications for plant performance. *Mol. Plant Microbe Interact.* 27: 349-363.

Baumert, A., Mock, H.-P., Schmidt, J., Herbers, K., Sonnewald, U. & Strack, D. (2001) Patterns of phenylpropanoids in non-inoculated and potato virus Y-inoculated leaves of transgenic tobacco plants expressing yeast-derived invertase. *Phytochemistry* 56: 535-541.

Bécard, G. & Piché, Y. (1989) Fungal growth stimulation by CO_2 and root exudates in vesicular-arbuscular mycorrhizal symbiosis. *Appl. Environ. Microbiol.* 55: 2320-2325.

Bécard, G., Douds, D.D. & Pfeffer, P.E. (1992) Extensive *in vitro* hyphal growth of vesicular-arbuscular mycorrhizal fungi in the presence of CO_2 and flavonols. *Appl. Environ. Microbiol.* 58: 821-825.

Bécard, G., Taylor, L.P., Douds, D.D., Pfeffer, P.E. & Doner, L.W. (1995) Flavonoids are not necessary plant signal compounds in arbuscular mycorrhizal symbioses. *Mol. Plant Microbe Interact.* 8: 252-258.

Bellincampi, D., Cervone, F. & Lionetti, V. (2014) Plant cell wall dynamics and wall-related susceptibility in plant-pathogen interactions. *Front. Plant Sci.* 5: 228.

Ben Amor, B., Shaw, S.L., Oldroyd, G.E.D., Maillet, F., Penmetsa, R.V., Cook, D., Long, S.R., Dénarié, J. & Gough, C. (2003) The *NFP* locus of *Medicago truncatula* controls an early step of Nod factor signal transduction upstream of a rapid calcium flux and root hair deformation. *Plant J.* 34: 495-506.

Berger, S., Papadopoulos, M., Schreiber, U., Kaiser, W. & Roitsch, T. (2004) Complex regulation of gene expression, photosynthesis and sugar levels by pathogen infection in tomato. *Physiol. Plant.* 122: 419-428.

Besserer, A., Bécard, G., Jauneau, A., Roux, C. & Séjalon-Delmas, N. (2008) GR24, a synthetic analog of strigolactones, stimulates the mitosis and growth of the arbuscular mycorrhizal fungus *Gigaspora rosea* by boosting its energy metabolism. *Plant Physiol.* 148: 402-413.

Besserer, A., Puech-Pagès, V., Kiefer, P., Gomez-Roldan, V., Jauneau, A., Roy, S., Portais, J.-C., Roux, C., Bécard, G. & Séjalon-Delmas, N. (2006) Strigolactones stimulate arbuscular mycorrhizal fungi by activating mitochondria. *PLoS Biol.* 4: e226.

Bienert, G., Thorsen, M., Schussler, M., Nilsson, H., Wagner, A., Tamas, M. & Jahn, T. (2008) A subgroup of plant aquaporins facilitate the bi-directional diffusion of $As(OH)_3$ and $Sb(OH)_3$ across membranes. *BMC Biol.* 6: 26.

Bienert, G.P. & Chaumont, F. (2014) Aquaporin-facilitated transmembrane diffusion of hydrogen peroxide. *Biochim. Biophys. Acta* 1840: 1596-1604.

Bitterlich, M., Krügel, U., Boldt-Burisch, K., Franken, P. & Kühn, C. (2014) The sucrose transporter SlSUT2 from tomato interacts with brassinosteroid functioning and affects arbuscular mycorrhiza formation. *Plant J.* 78: 877-889.

Blee, K.A. & Anderson, A.J. (1998) Regulation of arbuscule formation by carbon in the plant. *Plant J.* 16: 523-530.

Blee, K.A. & Anderson, A.J. (2002) Transcripts for genes encoding soluble acid invertase and sucrose synthase accumulate in root tip and cortical cells containing mycorrhizal arbuscules. *Plant Mol. Biol.* 50: 197-211.

Boldt, K., Pörs, Y., Haupt, B., Bitterlich, M., Kühn, C., Grimm, B. & Franken, P. (2011) Photochemical processes, carbon assimilation and RNA accumulation of sucrose transporter genes in tomato arbuscular mycorrhiza. *J. Plant Physiol.* 168: 1256-1263.

Bonfante-Fasolo, P., Tamagnone, L., Peretto, R., Esquerré-Tugayé, M.T., Mazau, D., Mosiniak, M. & Vian, B. (1991) Immunocytochemical location of hydroxyproline rich glycoproteins at the interface between a mycorrhizal fungus and its host plants. *Protoplasma* 165: 127-138.

Bonfante-Fasolo, P., Vian, B., Perotto, S., Faccio, A. & Knox, J.P. (1990) Cellulose and pectin localization in roots of mycorrhizal *Allium porrum*: labelling continuity between host cell wall and interfacial material. *Planta* 180: 537-547.

Bonfante, P. & Genre, A. (2008) Plants and arbuscular mycorrhizal fungi: an evolutionary-developmental perspective. *Trends Plant Sci.* 13: 492-498.

Bonfante, P., Genre, A., Faccio, A., Martini, I., Schauser, L., Stougaard, J., Webb, J. & Parniske, M. (2000) The *Lotus japonicus LjSym4* gene is required for the successful symbiotic infection of root epidermal cells. *Mol. Plant Microbe Interact.* 13: 1109-1120.

Bonfig, K.B., Schreiber, U., Gabler, A., Roitsch, T. & Berger, S. (2006) Infection with virulent and avirulent *P. syringae* strains differentially affects photosynthesis and sink metabolism in *Arabidopsis* leaves. *Planta* 225: 1-12.

Brewer, P.B., Koltai, H. & Beveridge, C.A. (2013) Diverse roles of strigolactones in plant development. *Mol. Plant* 6: 18-28.

Brewin, N. (2004) Plant cell wall remodelling in the rhizobium-legume symbiosis. *Crit. Rev. Plant Sci.* 23: 293-316.

Broghammer, A., Krusell, L., Blaise, M., Sauer, J., Sullivan, J.T., Maolanon, N., Vinther, M., Lorentzen, A., Madsen, E.B., Jensen, K.J., Roepstorff, P., Thirup, S., Ronson, C.W., Thygesen, M.B. & Stougaard, J. (2012) Legume receptors perceive the rhizobial lipochitin oligosaccharide signal molecules by direct binding. *Proc. Natl. Acad. Sci. USA* 109: 13859-13864.

Brundrett, M.C. (2002) Coevolution of roots and mycorrhizas of land plants. *New Phytol.* 154: 275-304.

Bucher, M. (2007) Functional biology of plant phosphate uptake at root and mycorrhiza interfaces. *New Phytol.* 173: 11-26.

Bücking, H. & Shachar-Hill, Y. (2005) Phosphate uptake, transport and transfer by the arbuscular mycorrhizal fungus *Glomus intraradices* is stimulated by increased carbohydrate availability. *New Phytol.* 165: 899-912.

Bücking, H., Abubaker, J., Govindarajulu, M., Tala, M., Pfeffer, P.E., Nagahashi, G., Lammers, P. & Shachar-Hill, Y. (2008) Root exudates stimulate the uptake and metabolism of organic carbon in germinating spores of *Glomus intraradices*. *New Phytol.* 180: 684-695.

Buee, M., Rossignol, M., Jauneau, A., Ranjeva, R. & Bécard, G. (2000) The pre-symbiotic growth of arbuscular mycorrhizal fungi is induced by a branching factor partially purified from plant root exudates. *Mol. Plant Microbe Interact.* 13: 693-698.

Caba, J.M., Centeno, M.L., Fernández, B., Gresshoff, P.M. & Ligero, F. (2000) Inoculation and nitrate alter phytohormone levels in soybean roots: differences between a supernodulating mutant and the wild type. *Planta* 211: 98-104.

Cabello-Hurtado, F. & Ramos, J. (2004) Isolation and functional analysis of the glycerol permease activity of two new nodulin-like intrinsic proteins from salt stressed roots of the halophyte *Atriplex nummularia*. *Plant Sci.* 166: 633-640.

Caetano-Anolles, G. & Gresshoff, P.M. (1990) Early induction of feedback regulatory responses governing nodulation in soybean. *Plant Sci.* 71: 69.

Campalans, A., Kondorosi, A. & Crespi, M. (2004) *Enod40*, a short open reading frame-containing mRNA, induces cytoplasmic localization of a nuclear RNA binding protein in *Medicago truncatula*. *Plant Cell* 16: 1047-1059.

Capoen, W., Sun, J., Wysham, D., Otegui, M.S., Venkateshwaran, M., Hirsch, S., Miwa, H., Downie, J.A., Morris, R.J., Ané, J.-M. & Oldroyd, G.E.D. (2011) Nuclear membranes control symbiotic calcium signaling of legumes. *Proc. Natl. Acad. Sci. USA* 108: 14348-14353.

Carroll, B.J., McNeil, D.L. & Gresshoff, P.M. (1985) A supernodulation and nitrate-tolerant symbiotic (nts) soybean mutant. *Plant Physiol.* 78: 34-40.

Catford, J.G., Staehelin, C., Lerat, S., Piche, Y. & Vierheilig, H. (2003) Suppression of arbuscular mycorrhizal colonization and nodulation in split-root systems of alfalfa after pre-inoculation and treatment with Nod factors. *J. Exp. Bot.* 54: 1481-1487.

Catford, J.G., Staehelin, C., Larose, G., Piché, Y. & Vierheilig, H. (2006) Systemically suppressed isoflavonoids and their stimulating effects on nodulation and mycorrhization in alfalfa split-root systems. *Plant Soil* 285: 257-266.

Catoira, R., Galera, C., de Billy, F., Penmetsa, R.V., Journet, E.-P., Maillet, F., Rosenberg, C., Cook, D., Gough, C. & Denarie, J. (2000) Four genes of

Medicago truncatula controlling components of a Nod factor transduction pathway. *Plant Cell* 12: 1647-1666.

Cerri, M.R., Frances, L., Laloum, T., Auriac, M.-C., Niebel, A., Oldroyd, G.E.D., Barker, D.G., Fournier, J. & de Carvalho-Niebel, F. (2012) *Medicago truncatula* ERN transcription factors: regulatory interplay with NSP1/NSP2 GRAS factors and expression dynamics throughout rhizobial infection. *Plant Physiol.* 160: 2155-2172.

Cesco, S., Mimmo, T., Tonon, G., Tomasi, N., Pinton, R., Terzano, R., Neumann, G., Weisskopf, L., Renella, G., Landi, L. & Nannipieri, P. (2012) Plant-borne flavonoids released into the rhizosphere: impact on soil bio-activities related to plant nutrition. A review. *Biol. Fert. Soils* 48: 123-149.

Chabaud, M., Genre, A., Sieberer, B.J., Faccio, A., Fournier, J., Novero, M., Barker, D.G. & Bonfante, P. (2011) Arbuscular mycorrhizal hyphopodia and germinated spore exudates trigger Ca^{2+} spiking in the legume and nonlegume root epidermis. *New Phytol.* 189: 347-355.

Chabaud, M., Venard, C., Defaux-Petras, A., Bécard, G. & Barker, D.G. (2002) Targeted inoculation of *Medicago truncatula in vitro* root cultures reveals *MtENOD11* expression during early stages of infection by arbuscular mycorrhizal fungi. *New Phytol.* 156: 265-273.

Charpentier, M., Bredemeier, R., Wanner, G., Takeda, N., Schleiff, E. & Parniske, M. (2008) *Lotus japonicus* CASTOR and POLLUX are ion channels essential for perinuclear calcium spiking in legume root endosymbiosis. *Plant Cell* 20: 3467-3479.

Charron, D., Pingret, J.-L., Chabaud, M., Journet, E.-P. & Barker, D.G. (2004) Pharmacological evidence that multiple phospholipid signaling pathways link rhizobium nodulation factor perception in *Medicago truncatula* root hairs to intracellular responses, including Ca^{2+} spiking and specific *ENOD* gene expression. *Plant Physiol.* 136: 3582-3593.

Clark, S.E. (2001) Cell signalling at the shoot meristem. *Nat. Rev. Mol. Cell Biol.* 2: 276-284.

Combier, J.-P., Frugier, F., de Billy, F., Boualem, A., El-Yahyaoui, F., Moreau, S., Vernié, T., Ott, T., Gamas, P., Crespi, M. & Niebel, A. (2006) *MtHAP2-1* is a key transcriptional regulator of symbiotic nodule development regulated by microRNA169 in *Medicago truncatula. Genes Dev.* 20: 3084-3088.

Combier, J.-P., de Billy, F., Gamas, P., Niebel, A. & Rivas, S. (2008a) *Trans*-regulation of the expression of the transcription factor *MtHAP2-1* by a uORF controls root nodule development. *Genes Dev.* 22: 1549-1559.

Combier, J.-P., Küster, H., Journet, E.-P., Hohnjec, N., Gamas, P. & Niebel, A. (2008b) Evidence for the involvement in nodulation of the two small putative regulatory peptide-encoding genes MtRALFL1 and MtDVL1. *Mol. Plant Microbe Interact.* 21: 1118-1127.

Cook, C.E., Whichard, L.P., Turner, B., Wall, M.E. & Egley, G.H. (1966) Germination of witchweed (*Striga lutea* Lour.): Isolation and properties of a potent stimulant. *Science* 154: 1189-1190.

Cook, C.E., Whichard, L.P., Wall, M., Egley, G.H., Coggon, P., Luhan, P.A. & McPhail, A.T. (1972) Germination stimulants. II. Structure of strigol, a potent seed germination stimulant for witchweed (*Striga lutea*). *J. Am. Chem. Soc.* 94: 6198-6199.

Corradi, N. & Bonfante, P. (2012) The arbuscular mycorrhizal symbiosis: origin and evolution of a beneficial plant infection. *PLoS Pathog.* 8: e1002600.

Cosgrove, D.J. (2000) Loosening of plant cell walls by expansins. *Nature* 407: 321-326.

Cox, G.U.Y. & Tinker, P.B. (1976) Translocation and transfer of nutrients in vesicular-arbuscular mycorrhizas. *New Phytol.* 77: 371-378.

Czaja, L.F., Hogekamp, C., Lamm, P., Maillet, F., Martinez, E.A., Samain, E., Dénarié, J., Küster, H. & Hohnjec, N. (2012) Transcriptional responses toward diffusible signals from symbiotic microbes reveal *MtNFP*- and *MtDMI3*-dependent reprogramming of host gene expression by arbuscular mycorrhizal fungal lipochitooligosaccharides. *Plant Physiol.* 159: 1671-1685.

D'Haeze, W. & Holsters, M. (2002) Nod factor structures, responses, and perception during initiation of nodule development. *Glycobiology* 12: 79R-105R.

de Carvalho-Niebel, F., Timmers, A.C.J., Chabaud, M., Defaux-Petras, A. & Barker, D.G. (2002) The Nod factor-elicited annexin MtAnn1 is preferentially localised at the nuclear periphery in symbiotically activated root tissues of *Medicago truncatula*. *Plant J.* 32: 343-352.

de Los Santos, R.T., Vierheilig, H., Ocampo, J.A. & García Garrido, J.M. (2011) Altered pattern of arbuscular mycorrhizal formation in tomato ethylene mutants. *Plant Signal. Behav.* 6: 755-758.

De Mita, S., Streng, A., Bisseling, T. & Geurts, R. (2014) Evolution of a symbiotic receptor through gene duplications in the legume–rhizobium mutualism. *New Phytol.* 201: 961-972.

Dean, R.M., Rivers, R.L., Zeidel, M.L. & Roberts, D.M. (1998) Purification and functional reconstitution of soybean Nodulin 26. An aquaporin with water and glycerol transport properties. *Biochemistry* 38: 347-353.

Dehne, H.W. (1986) Influence of VA mycorrhizae on host plant physiology. In: *Physiological and Genetical Aspects of Mycorrhizae*. V. Gianinazzi-Pearson & S. Gianinazzi (Eds.) Paris: Institut National de la Recherche Agronomique. 431-435.

Delves, A.C., Mathews, A., Day, D.A., Carter, A.S., Carroll, B.J. & Gresshoff, P.M. (1986) Regulation of the soybean-rhizobium nodule symbiosis by shoot and root factors. *Plant Physiol.* 82: 588-590.

Demchenko, K., Winzer, T., Stougaard, J., Parniske, M. & Pawlowski, K. (2004) Distinct roles of *Lotus japonicus SYMRK* and *SYM15* in root colonization and arbuscule formation. *New Phytol.* 163: 381-392.

Dermatsev, V., Weingarten-Baror, C., Resnick, N., Gadkar, V., Wininger, S., Kolotilin, I., Mayzlish-Gati, E., Zilberstein, A., Koltai, H. & Kapulnik, Y. (2010) Microarray analysis and functional tests suggest the involvement of expansins in the early stages of symbiosis of the arbuscular mycorrhizal fungus *Glomus intraradices* on tomato (*Solanum lycopersicum*). *Mol. Plant Pathol.* 11: 121-135.

Den Hartog, M., Musgrave, A. & Munnik, T. (2001) Nod factor-induced phosphatidic acid and diacylglycerol pyrophosphate formation: a role for phospholipase C and D in root hair deformation. *Plant J.* 25: 55-65.

Devers, E.A., Branscheid, A., May, P. & Krajinski, F. (2011) Stars and symbiosis: microRNA- and microRNA*-mediated transcript cleavage involved in arbuscular mycorrhizal symbiosis. *Plant Physiol.* 156: 1990-2010.

Devers, E.A., Teply, J., Reinert, A., Gaude, N. & Krajinski, F. (2013) An endogenous artificial microRNA system for unraveling the function of root endosymbioses related genes in *Medicago truncatula*. *BMC Plant Biol.* 13: 82.

Dickson, S. (2004) The *Arum–Paris* continuum of mycorrhizal symbioses. *New Phytol.* 163: 187-200.

Dickson, S., Smith, F.A. & Smith, S.E. (2007) Structural differences in arbuscular mycorrhizal symbioses: more than 100 years after Gallaud, where next? *Mycorrhiza* 17: 375-393.

Dietz, K.-J., Vogel, M. & Viehhauser, A. (2010) AP2/EREBP transcription factors are part of gene regulatory networks and integrate metabolic, hormonal and environmental signals in stress acclimation and retrograde signalling. *Protoplasma* 245: 3-14.

Doidy, J., van Tuinen, D., Lamotte, O., Corneillat, M., Alcaraz, G. & Wipf, D. (2012) The *Medicago truncatula* sucrose transporter family: Characterization and implication of key members in carbon partitioning towards arbuscular mycorrhizal fungi. *Mol. Plant* 5: 1346-1358.

Douds, D.D. (1994) Relationship between hyphal and arbuscular colonization and sporulation in a mycorrhiza of *Paspalum notatum* Flugge. *New Phytol.* 126: 233-237.

Douds, D.D., Pfeffer, P.E. & Shachar-Hill, Y. (2000) Carbon partitioning, cost, and metabolism of arbuscular mycorrhizas. In: *Arbuscular Mycorrhizas: Physiology and Function.* Y. Kapulnik & D.D. Douds (Eds.) Dordrecht: Kluwer Academic Publishers. 107-129.

Drissner, D., Kunze, G., Callewaert, N., Gehrig, P., Tamasloukht, M.B., Boller, T., Felix, G., Amrhein, N. & Bucher, M. (2007) Lyso-phosphatidylcholine is a signal in the arbuscular mycorrhizal symbiosis. *Science* 318: 265-268.

Djordjevic, M.A., Oakes, M., Li, D.X., Hwang, C.H., Hocart, C.H. & Gresshoff, P.M. (2007) The *Glycine max* xylem sap and apoplast proteome. *J. Proteome Res.* 6: 3771-3779.

El-Mesbahi, M.N., Azcón, R., Ruiz-Lozano, J.M. & Aroca, R. (2012) Plant potassium content modifies the effects of arbuscular mycorrhizal symbiosis on root hydraulic properties in maize plants. *Mycorrhiza* 22: 555-564.

Elias, K.S. & Safir, G.R. (1987) Hyphal elongation of *Glomus fasciculatus* in response to root exudates. *Appl. Environ. Microbiol.* 53: 1928-1933.

Engstrom, E.M., Ehrhardt, D.W., Mitra, R.M. & Long, S.R. (2002) Pharmacological analysis of Nod factor-induced calcium spiking in *Medicago truncatula*. Evidence for the requirement of type IIA calcium pumps and phosphoinositide signaling. *Plant Physiol.* 128: 1390-1401.

Essmann, J., Schmitz-Thom, I., Schön, H., Sonnewald, S., Weis, E. & Scharte, J. (2008) RNA interference-mediated repression of cell wall invertase impairs defense in source leaves of tobacco. *Plant Physiol.* 147: 1288-1299.

Feddermann, N., Duvvuru Muni, R.R., Zeier, T., Stuurman, J., Ercolin, F., Schorderet, M. & Reinhardt, D. (2010) The *PAM1* gene of petunia, required for intracellular accommodation and morphogenesis of arbuscular mycorrhizal fungi, encodes a homologue of VAPYRIN. *Plant J.* 64: 470-481.

Fellbaum, C.R., Gachomo, E.W., Beesetty, Y., Choudhari, S., Strahan, G.D., Pfeffer, P.E., Kiers, E.T. & Bücking, H. (2012) Carbon availability triggers fungal nitrogen uptake and transport in arbuscular mycorrhizal symbiosis. *Proc. Natl. Acad. Sci. USA* 109: 2666-2671.

Fellbaum, C.R., Mensah, J.A., Cloos, A.J., Strahan, G.E., Pfeffer, P.E., Kiers, E.T. & Bücking, H. (2014) Fungal nutrient allocation in common mycorrhizal networks is regulated by the carbon source strength of individual host plants. *New Phytol.* 203: 646-656.

Felle, H.H. (1998) The apoplastic pH of the *Zea mays* root cortex as measured with pH-sensitive microelectrodes: aspects of regulation. *J. Exp. Bot.* 49: 987-995.

Ferguson, B.J., Indrasumunar, A., Hayashi, S., Lin, M.-H., Lin, Y.-H., Reid, D.E. & Gresshoff, P.M. (2010) Molecular analysis of legume nodule development and autoregulation. *J. Integr. Plant Biol.* 52: 61-76.

Fester, T., Strack, D. & Hause, B. (2001) Reorganization of tobacco root plastids during arbuscule development. *Planta* 213: 864-868.

Floß, D.S., Hause, B., Lange, P.R., Küster, H., Strack, D. & Walter, M.H. (2008a) Knock-down of the MEP pathway isogene *1-deoxy-d-xylulose 5-phosphate synthase 2* inhibits formation of arbuscular mycorrhiza-induced apocarotenoids, and abolishes normal expression of mycorrhiza-specific plant marker genes. *Plant J.* 56: 86-100.

Floß, D.S., Levy, J.G., Lévesque-Tremblay, V., Pumplin, N. & Harrison, M.J. (2013) DELLA proteins regulate arbuscule formation in arbuscular mycorrhizal symbiosis. *Proc. Natl. Acad. Sci. USA* 110: E5025-E5034.

Floß, D.S., Schliemann, W., Schmidt, J., Strack, D. & Walter, M.H. (2008b) RNA interference-mediated repression of *MtCCD1* in mycorrhizal roots of *Medicago truncatula* causes accumulation of C_{27} apocarotenoids, shedding light on the functional role of CCD1. *Plant Physiol.* 148: 1267-1282.

Foo, E. (2013) Auxin influences strigolactones in pea mycorrhizal symbiosis. *J. Plant Physiol.* 170: 523-528.

Foo, E. & Davies, N. (2011) Strigolactones promote nodulation in pea. *Planta* 234: 1073-1081.

Foo, E., Ross, J.J., Jones, W.T. & Reid, J.B. (2013a) Plant hormones in arbuscular mycorrhizal symbioses: an emerging role for gibberellins. *Ann. Bot.* 111: 769-779.

Foo, E., Yoneyama, K., Hugill, C.J., Quittenden, L.J. & Reid, J.B. (2013b) Strigolactones and the regulation of pea symbioses in response to nitrate and phosphate deficiency. *Mol. Plant* 6: 76-87.

Fortin, J.A., Bécard, G., Declerck, S., Dalpé, Y., St-Arnaud, M., Coughlan, A.P. & Piché, Y. (2002) Arbuscular mycorrhiza on root-organ cultures. *Can. J. Bot.* 80: 1-20.

Fotopoulos, V. (2005) Plant invertases: structure, function and regulation of a diverse enzyme family. *J. Biol. Res.* 4: 127-137.

Frank, B. (1885) Über die auf Wurzelsymbiose beruhende Ernährung gewisser Bäume durch unterirdische Pilze. Berichte der deutschen botanischen *Gesellschaft* 3: 128-145.

Gadkar, V., David-Schwartz, R., Nagahashi, G., Douds, D.D., Wininger, S. & Kapulnik, Y. (2003) Root exudate of *pmi* tomato mutant M161 reduces AM fungal proliferation in vitro. *FEMS Microbiol. Lett.* 223: 193-198.

Gao, D., Knight, M.R., Trewavas, A.J., Sattelmacher, B. & Plieth, C. (2004) Self-reporting Arabidopsis expressing pH and [Ca^{2+}] indicators unveil ion dynamics in the cytoplasm and in the apoplast under abiotic stress. *Plant Physiol.* 134: 898-908.

Gao, X., Lu, X., Wu, M., Zhang, H., Pan, R., Tian, J., Li, S. & Liao, H. (2012) Co-inoculation with rhizobia and AMF inhibited soybean red crown rot: from field study to plant defense-related gene expression analysis. *PLoS ONE.* 7: e33977.

García-Rodríguez, S., Azcón-Aguilar, C. & Ferrol, N. (2007) Transcriptional regulation of host enzymes involved in the cleavage of sucrose during arbuscular mycorrhizal symbiosis. *Physiol. Plant.* 129: 737-746.

Garrido, J.M.G., Morcillo, R.J.L., Rodríguez, J.Á.M. & Bote, J.A.O. (2010) Variations in the mycorrhization characteristics in roots of wild-type and ABA-deficient tomato are accompanied by specific transcriptomic alterations. *Mol. Plant Microbe Interact.* 23: 651-664.

Gaspar, M.L., Pollero, R.J. & Cabello, M.N. (1994) Triacylglycerol consumption during spore germination of vesicular-arbuscular mycorrhizal fungi. *J. Am. Oil Chem. Soc.* 71: 449-452.

Gaude, N., Bortfeld, S., Duensing, N., Lohse, M. & Krajinski, F. (2012) Arbuscule-containing and non-colonized cortical cells of mycorrhizal roots undergo extensive and specific reprogramming during arbuscular mycorrhizal development. *Plant J.* 69: 510-528.

Genre, A., Chabaud, M., Balzergue, C., Puech-Pagès, V., Novero, M., Rey, T., Fournier, J., Rochange, S., Bécard, G., Bonfante, P. & Barker, D.G. (2013) Short-chain chitin oligomers from arbuscular mycorrhizal fungi trigger nuclear Ca^{2+} spiking in *Medicago truncatula* roots and their production is enhanced by strigolactone. *New Phytol.* 198: 190-202.Genre, A., Chabaud, M., Faccio, A., Barker, D.G. & Bonfante, P. (2008)

Prepenetration apparatus assembly precedes and predicts the colonization patterns of arbuscular mycorrhizal fungi within the root cortex of both *Medicago truncatula* and *Daucus carota*. *Plant Cell* 20: 1407-1420.

Genre, A., Chabaud, M., Timmers, T., Bonfante, P. & Barker, D.G. (2005) Arbuscular mycorrhizal fungi elicit a novel intracellular apparatus in *Medicago truncatula* root epidermal cells before infection. *Plant Cell* 17: 3489-3499.

Genre, A., Ivanov, S., Fendrych, M., Faccio, A., Žárský, V., Bisseling, T. & Bonfante, P. (2012) Multiple exocytotic markers accumulate at the sites of perifungal membrane biogenesis in arbuscular mycorrhizas. *Plant Cell Physiol.* 53: 244-255.

Genre, A., Ortu, G., Bertoldo, C., Martino, E. & Bonfante, P. (2009) Biotic and abiotic stimulation of root epidermal cells reveals common and specific responses to arbuscular mycorrhizal fungi. *Plant Physiol.* 149: 1424-1434.

Gherbi, H., Markmann, K., Svistoonoff, S., Estevan, J., Autran, D., Giczey, G., Auguy, F., Péret, B., Laplaze, L., Franche, C., Parniske, M. & Bogusz, D. (2008) SymRK defines a common genetic basis for plant root endosymbioses with arbuscular mycorrhiza fungi, rhizobia, and *Frankia* bacteria. *Proc. Natl. Acad. Sci. USA* 105: 4928-4932.

Gianinazzi-Pearson, V. & Gianinazzi, S. (1992) Influence of intergeneric grafts between host and non-host legumes on formation of vesicular-arbuscular mycorrhiza. *New Phytol.* 120: 505-508.

Gianinazzi-Pearson, V., Arnould, C., Oufattole, M., Arango, M. & Gianinazzi, S. (2000) Differential activation of H^+-ATPase genes by an arbuscular mycorrhizal fungus in root cells of transgenic tobacco. *Planta* 211: 609-613.

Gianinazzi-Pearson, V., Smith, S.E., Gianinazzi, S. & Smith, F.A. (1991) Enzymatic studies on the metabolism of vesicular-arbuscular mycorrhizas. V. Is H^+-ATPase a component of ATP-hydrolysing enzyme activities in plant-fungus interfaces? *New Phytol.* 117: 61-74.

Gianinazzi, S., Gollotte, A., Binet, M.-N., van Tuinen, D., Redecker, D. & Wipf, D. (2010) Agroecology: the key role of arbuscular mycorrhizas in ecosystem services. *Mycorrhiza* 20: 519-530.

Giovannetti, M., Balestrini, R., Volpe, V., Guether, M., Straub, D., Costa, A., Ludewig, U. & Bonfante, P. (2012) Two putative-aquaporin genes are differentially expressed during arbuscular mycorrhizal symbiosis in *Lotus japonicus*. *BMC Plant Biol.* 12: 186.

Glassop, D., Smith, S. & Smith, F. (2005) Cereal phosphate transporters associated with the mycorrhizal pathway of phosphate uptake into roots. *Planta* 222: 688-698.

Gleason, C., Chaudhuri, S., Yang, T., Munoz, A., Poovaiah, B.W. & Oldroyd, G.E.D. (2006) Nodulation independent of rhizobia induced by a calcium-activated kinase lacking autoinhibition. *Nature* 441: 1149-1152.

Gobbato, E., Marsh, John F., Vernié, T., Wang, E., Maillet, F., Kim, J., Miller, J.B., Sun, J., Bano, S.A., Ratet, P., Mysore, Kirankumar S., Dénarié, J., Schultze, M. & Oldroyd, Giles E.D. (2012) A GRAS-type transcription factor with a specific function in mycorrhizal signaling. *Curr. Biol.* 22: 2236-2241.

Gobbato, E., Wang, E., Higgins, G., Bano, S., Henry, C., Schultze, M. & Oldroyd, G.E. (2013) *RAM1* and *RAM2* function and expression during arbuscular mycorrhizal symbiosis and *Aphanomyces euteiches* colonization. *Plant Signal. Behav.* 8: e26049.

Godt, D.E. & Roitsch, T. (1997) Regulation and tissue-specific distribution of mRNAs for three extracellular invertase isoenzymes of tomato suggests an important function in establishing and maintaining sink metabolism. *Plant Physiol.* 115: 273-282.

Goetz, M., Godt, D.E. & Roitsch, T. (2000) Tissue-specific induction of the mRNA for an extracellular invertase isoenzyme of tomato by brassinosteroids suggests a role for steroid hormones in assimilate partitioning. *Plant J.* 22: 515-522.

Gomez, S.K., Javot, H., Deewatthanawong, P., Torres-Jerez, I., Tang, Y., Blancaflor, E., Udvardi, M. & Harrison, M.J. (2009) *Medicago truncatula* and *Glomus intraradices* gene expression in cortical cells harboring arbuscules in the arbuscular mycorrhizal symbiosis. *BMC Plant Biol.* 9: 10.

Gough, C. & Cullimore, J. (2011) Lipo-chitooligosaccharide signaling in endosymbiotic plant-microbe interactions. *Mol. Plant. Microbe Interact.* 24: 867-878.

Graham, J.H. (1982) Effect of citrus root exudates on germination of chlamydospores of the vesicular-arbuscular mycorrhizal fungus, *Glomus epigaeum*. *Mycologia 74*: 831-835.

Graham, J.H. (2000) Assessing costs of arbuscular mycorrhizal symbiosis agroecosystems fungi. In: *Current Advances in Mycorrhizae Research.* G.K. Podila & D.D. Douds (Eds.) St. Paul: APS Press. 127-140.

Grignon, C. & Sentenac, H. (1991) pH and ionic conditions in the apoplast. *Annu Rev Plant Physiol. Plant Mol. Biol.* 42: 103-128.

Groth, M., Takeda, N., Perry, J., Uchida, H., Dräxl, S., Brachmann, A., Sato, S., Tabata, S., Kawaguchi, M., Wang, T.L. & Parniske, M. (2010) *NENA*, a *Lotus japonicus* homolog of *Sec13*, is required for rhizodermal infection by arbuscular mycorrhiza fungi and rhizobia but dispensable for cortical endosymbiotic development. *Plant Cell* 22: 2509-2526.

Guenther, J.F., Chanmanivone, N., Galetovic, M.P., Wallace, I.S., Cobb, J.A. & Roberts, D.M. (2003) Phosphorylation of soybean Nodulin 26 on serine 262 enhances water permeability and is regulated developmentally and by osmotic signals. *Plant Cell* 15: 981-991.

Guether, M., Balestrini, R., Hannah, M., He, J., Udvardi, M.K. & Bonfante, P. (2009a) Genome-wide reprogramming of regulatory networks, transport, cell wall and membrane biogenesis during arbuscular mycorrhizal symbiosis in *Lotus japonicus*. *New Phytol.* 182: 200-212.

Guether, M., Neuhäuser, B., Balestrini, R., Dynowski, M., Ludewig, U. & Bonfante, P. (2009b) A mycorrhizal-specific ammonium transporter from *Lotus japonicus* acquires nitrogen released by arbuscular mycorrhizal fungi. *Plant Physiol.* 150: 73-83.

Gupta Sood, S. (2003) Chemotactic response of plant-growth-promoting bacteria towards roots of vesicular-arbuscular mycorrhizal tomato plants. *FEMS Microbiol. Ecol.* 45: 219-227.

Gutjahr, C., Banba, M., Croset, V., An, K., Miyao, A., An, G., Hirochika, H., Imaizumi-Anraku, H. & Paszkowski, U. (2008) Arbuscular mycorrhiza–specific signaling in rice transcends the common symbiosis signaling pathway. *Plant Cell* 20: 2989-3005.

Gutjahr, C., Radovanovic, D., Geoffroy, J., Zhang, Q., Siegler, H., Chiapello, M., Casieri, L., An, K., An, G., Guiderdoni, E., Kumar, C.S., Sundaresan, V., Harrison, M.J. & Paszkowski, U. (2012) The half-size ABC transporters STR1 and STR2 are indispensable for mycorrhizal arbuscule formation in rice. *Plant J.* 69: 906-920.

Guttenberger, M. (2000) Arbuscules of vesicular-arbuscular mycorrhizal fungi inhabit an acidic compartment within plant roots. *Planta* 211: 299-304.

Hamacher, T., Becker, J., Gárdonyi, M., Hahn-Hägerdal, B. & Boles, E. (2002) Characterization of the xylose-transporting properties of yeast hexose transporters and their influence on xylose utilization. *Microbiology* 148: 2783-2788.

Hammer, E.C., Pallon, J., Wallander, H. & Olsson, P.A. (2011) Tit for tat? A mycorrhizal fungus accumulates phosphorus under low plant carbon availability. *FEMS Microbiol. Ecol.* 76: 236-244.

Hammond, J.P. & White, P.J. (2008) Sucrose transport in the phloem: integrating root responses to phosphorus starvation. *J. Exp. Bot.* 59: 93-109.

Hammond, J.P. & White, P.J. (2011) Sugar signaling in root responses to low phosphorus availability. *Plant Physiol.* 156: 1033-1040.

Hans, J., Hause, B., Strack, D. & Walter, M.H. (2004) Cloning, characterization, and immunolocalization of a mycorrhiza-inducible 1-deoxy-D-xylulose 5-phosphate reductoisomerase in arbuscule-containing cells of maize. *Plant Physiol.* 134: 614-624.

Harrier, L. & Millam, S. (2001) Biolistic transformation of arbuscular mycorrhizal fungi. *Mol. Biotechnol.* 18: 25-33.

Harrison, M. (1996) A sugar transporter from *Medicago truncatula*: altered expression pattern in roots during vesicular-arbuscular (VA) mycorrhizal associations. *Plant J.* 9: 491-503.

Harrison, M.J., Dewbre, G.R. & Liu, J. (2002) A phosphate transporter from *Medicago truncatula* involved in the acquisition of phosphate released by arbuscular mycorrhizal fungi. *Plant Cell* 14: 2413-2429.

Hassan, S. & Mathesius, U. (2012) The role of flavonoids in root–rhizosphere signalling: opportunities and challenges for improving plant–microbe interactions. *J. Exp. Bot.* 63: 3429-3444.

Hause, B. & Schaarschmidt, S. (2009) The role of jasmonates in mutualistic symbioses between plants and soil-born microorganisms. *Phytochemistry* 70: 1589-1599.

Hayashi, S., Reid, D.E., Lorenc, M.T., Stiller, J., Edwards, D., Gresshoff, P.M. & Ferguson, B.J. (2012) Transient Nod factor-dependent gene expression in the nodulation-competent zone of soybean (*Glycine max* [L.] Merr.) roots. *Plant Biotechnol. J.* 10: 995-1010.

Hayman, D.S. (1974) Plant growth responses to vesicular-arbuscular mycorrhiza. *New Phytol.* 73: 71-80.

Helber, N., Wippel, K., Sauer, N., Schaarschmidt, S., Hause, B. & Requena, N. (2011) A versatile monosaccharide transporter that operates in the arbuscular mycorrhizal fungus *Glomus* sp is crucial for the symbiotic relationship with plants. *Plant Cell* 23: 3812-3823.

Herbers, K., Meuwly, P., Frommer, W., Métraux, J.-P. & Sonnewald, U. (1996) Systemic acquired resistance mediated by the ectopic expression of invertase: possible hexose sensing in the secretory pathway. *Plant Cell* 8: 793-803.

Herbers, K., Takahata, Y., Melzer, M., Mock, H.-P., Hajirezaei, M. & Sonnewald, U. (2000) Regulation of carbohydrate partitioning during the interaction of potato virus Y with tobacco. *Mol. Plant Pathol.* 1: 51-59.

Herrera-Medina, M.J., Steinkellner, S., Vierheilig, H., Ocampo Bote, J.A. & Garcia Garrido, J.M. (2007) Abscisic acid determines arbuscule development and functionality in the tomato arbuscular mycorrhiza. *New Phytol.* 175: 554-564.

Hirsch, S., Kim, J., Muñoz, A., Heckmann, A.B., Downie, J.A. & Oldroyd, G.E.D. (2009) GRAS proteins form a DNA binding complex to induce gene expression during nodulation signaling in *Medicago truncatula*. *Plant Cell* 21: 545-557.

Hogekamp, C. & Küster, H. (2013) A roadmap of cell-type specific gene expression during sequential stages of the arbuscular mycorrhiza symbiosis. *BMC Genomics* 14: 306.

Hogekamp, C., Arndt, D., Pereira, P.A., Becker, J.D., Hohnjec, N. & Küster, H. (2011) Laser microdissection unravels cell-type-specific transcription in arbuscular mycorrhizal roots, including CAAT-box transcription factor gene expression correlating with fungal contact and spread. *Plant Physiol.* 157: 2023-2043.

Hohnjec, N., Perlick, A.M., Pühler, A. & Küster, H. (2003) The *Medicago truncatula* sucrose synthase gene *MtSucS1* is activated both in the infected region of root nodules and in the cortex of roots colonized by arbuscular mycorrhizal fungi. *Mol. Plant Microbe Interact.* 16: 903-915.

Horváth, B., Yeun, L.H., Domonkos, Á., Halász, G., Gobbato, E., Ayaydin, F., Miró, K., Hirsch, S., Sun, J., Tadege, M., Ratet, P., Mysore, K.S., Ané, J.-M., Oldroyd, G.E.D. & Kaló, P. (2011) *Medicago truncatula IPD3* is a member of the common symbiotic signaling pathway required for rhizobial and mycorrhizal symbioses. *Mol. Plant Microbe Interact.* 24: 1345-1358.

Hu, Z., Yan, H., Yang, J., Yamaguchi, S., Maekawa, M., Takamure, I., Tsutsumi, N., Kyozuka, J. & Nakazono, M. (2010) Strigolactones negatively regulate mesocotyl elongation in rice during germination and growth in darkness. *Plant Cell Physiol.* 51: 1136-1142

Humphreys, C.P., Franks, P.J., Rees, M., Bidartondo, M.I., Leake, J.R. & Beerling, D.J. (2010) Mutualistic mycorrhiza-like symbiosis in the most ancient group of land plants. *Nat. Commun.* 1: 103.

Imaizumi-Anraku, H., Takeda, N., Charpentier, M., Perry, J., Miwa, H., Umehara, Y., Kouchi, H., Murakami, Y., Mulder, L., Vickers, K., Pike, J., Allan Downie, J., Wang, T., Sato, S., Asamizu, E., Tabata, S., Yoshikawa,

M., Murooka, Y., Wu, G.-J., Kawaguchi, M., Kawasaki, S., Parniske, M. & Hayashi, M. (2005) Plastid proteins crucial for symbiotic fungal and bacterial entry into plant roots. *Nature* 433: 527-531.

Isayenkov, S., Mrosk, C., Stenzel, I., Strack, D. & Hause, B. (2005) Suppression of allene oxide cyclase in hairy roots of *Medicago truncatula* reduces jasmonate levels and the degree of mycorrhization with *Glomus intraradices*. *Plant Physiol.* 139: 1401-1410.

Ivanov, S., Fedorova, E.E., Limpens, E., De Mita, S., Genre, A., Bonfante, P. & Bisseling, T. (2012) Rhizobium–legume symbiosis shares an exocytotic pathway required for arbuscule formation. *Proc. Natl. Acad. Sci. USA* 109: 8316-8321.

Javot, H., Penmetsa, R.V., Terzaghi, N., Cook, D.R. & Harrison, M.J. (2007a) A *Medicago truncatula* phosphate transporter indispensable for the arbuscular mycorrhizal symbiosis. *Proc. Natl. Acad. Sci. USA* 30: 1720-1725.

Javot, H., Pumplin, N. & Harrison, M.J. (2007b) Phosphate in the arbuscular mycorrhizal symbiosis: transport properties and regulatory roles. *Plant Cell Environ.* 30: 310-322.

Johnson, D., Leake, J.R., Ostle, N., Ineson, P. & Read, D.J. (2002) *In situ* $^{13}CO_2$ pulse-labelling of upland grassland demonstrates a rapid pathway of carbon flux from arbuscular mycorrhizal mycelia to the soil. *New Phytol.* 153: 327-334.

Journet, E.-P., El-Gachtouli, N., Vernoud, V., de Billy, F., Pichon, M., Dedieu, A., Arnould, C., Morandi, D., Barker, D.G. & Gianinazzi-Pearson, V. (2001) *Medicago truncatula ENOD11*: a novel RPRP-encoding early nodulin gene expressed during mycorrhization in arbuscule-containing cells. *Mol. Plant Microbe Interact.* 14: 737-748.

Juszczuk, I., Wiktorowska, A., Malusá, E. & Rychter, A. (2004) Changes in the concentration of phenolic compounds and exudation induced by phosphate deficiency in bean plants (*Phaseolus vulgaris* L.). *Plant Soil* 267: 41-49.

Kálo, P., Gleason, C., Edwards, A., Marsh, J., Mitra, R.M., Hirsch, S., Jakab, J., Sims, S., Long, S.R., Rogers, J., Kiss, G.B., Downie, J.A. & Oldroyd, G.E.D. (2005) Nodulation signaling in legumes requires NSP2, a member of the GRAS family of transcriptional regulators. *Science* 308: 1786-1789.

Kamienski, F. (1881) Die Vegetationsorgane der *Monotropa hypopitys* L. *Bot. Zeitschr.* 39: 225-234.

Kanamori, N., Madsen, L.H., Radutoiu, S., Frantescu, M., Quistgaard, E.M.H., Miwa, H., Downie, J.A., James, E.K., Felle, H.H., Haaning, L.L., Jensen,

T.H., Sato, S., Nakamura, Y., Tabata, S., Sandal, N. & Stougaard, J. (2006) A nucleoporin is required for induction of Ca^{2+} spiking in legume nodule development and essential for rhizobial and fungal symbiosis. *Proc. Natl. Acad. Sci. USA* 103: 359-364.

Kang, H., Chu, X., Wang, C., Xiao, A., Zhu, H., Yuan, S., Yang, Z., Ke, D., Xiao, S., Hong, Z. & Zhang, Z. (2014) A MYB coiled-coil transcription factor interacts with NSP2 and is involved in nodulation in *Lotus japonicus*. *New Phytol.* 201: 837-849.

Kang, H., Zhu, H., Chu, X., Yang, Z., Yuan, S., Yu, D., Wang, C., Hong, Z. & Zhang, Z. (2011) A novel interaction between CCaMK and a protein containing the Scythe_N ubiquitin-like domain in *Lotus japonicus*. *Plant Physiol.* 155: 1312-1324.

Karandashov, V., Nagy, R., Wegmüller, S., Amrhein, N. & Bucher, M. (2004) Evolutionary conservation of a phosphate transporter in the arbuscular mycorrhizal symbiosis. *Proc. Natl. Acad. Sci. USA* 101: 6285–6290.

Kenjo, T., Yamaya, H. & Arima, Y. (2010) Shoot-synthesized nodulation-restricting substances of wild-type soybean present in two different high-performance liquid chromatography peaks of the ethanol-soluble medium-polarity fraction. *Soil Sci. Plant Nutr.* 56: 399-406.

Kevei, Z., Lougnon, G., Mergaert, P., Horváth, G.V., Kereszt, A., Jayaraman, D., Zaman, N., Marcel, F., Regulski, K., Kiss, G.B., Kondorosi, A., Endre, G., Kondorosi, E. & Ané, J.-M. (2007) 3-Hydroxy-3-methylglutaryl coenzyme A reductase1 interacts with NORK and is crucial for nodulation in *Medicago truncatula*. *Plant Cell* 19: 3974-3989.

Kiers, E.T., Duhamel, M., Beesetty, Y., Mensah, J.A., Franken, O., Verbruggen, E., Fellbaum, C.R., Kowalchuk, G.A., Hart, M.M., Bago, A., Palmer, T.M., West, S.A., Vandenkoornhuyse, P., Jansa, J. & Bücking, H. (2011) Reciprocal rewards stabilize cooperation in the mycorrhizal symbiosis. *Science* 333: 880-882.

Kinkema, M. & Gresshoff, P.M. (2008) Investigation of downstream signals of the soybean autoregulation of nodulation receptor kinase GmNARK. *Mol. Plant Microbe Interact.* 21: 1337-1348.

Kistner, C., Winzer, T., Pitzschke, A., Mulder, L., Sato, S., Kaneko, T., Tabata, S., Sandal, N., Stougaard, J., Webb, K.J., Szczyglowski, K. & Parniske, M. (2005) Seven *Lotus japonicus* genes required for transcriptional reprogramming of the root during fungal and bacterial symbiosis. *Plant Cell* 17: 2217-2229.

Kloppholz, S., Kuhn, H. & Requena, N. (2011) A secreted fungal effector of *Glomus intraradices* promotes symbiotic biotrophy. *Curr. Biol.* 21: 1204-1209.

Kobae, Y., Tamura, Y., Takai, S., Banba, M. & Hata, S. (2010) Localized expression of arbuscular mycorrhiza-inducible ammonium transporters in soybean. *Plant Cell Physiol.* 51: 1411-1415.

Koegel, S., Ait Lahmidi, N., Arnould, C., Chatagnier, O., Walder, F., Ineichen, K., Boller, T., Wipf, D., Wiemken, A. & Courty, P.-E. (2013) The family of ammonium transporters (AMT) in *Sorghum bicolor*: two AMT members are induced locally, but not systemically in roots colonized by arbuscular mycorrhizal fungi. *New Phytol.* 198: 853-865.

Koske, R.E. (1981) Multiple germination by spores of *Gigaspora gigantea. Trans. Br. Mycol. Soc.* 76: 328-330.

Koske, R.E. (1982) Evidence for a volatile attractant from plant roots affecting germ tubes of a VA mycorrhizal fungus. *Trans. Br. Mycol.* 79: 305-310.

Kosslak, R.M. & Bohlool, B.B. (1984) Suppression of nodule development of one side of a split-root system of soybeans caused by prior inoculation of the other side. *Plant Physiol.* 75: 125-130.

Kosuta, S., Chabaud, M., Lougnon, G., Gough, C., Denarie, J., Barker, D.G. & Bécard, G. (2003) A diffusible factor from arbuscular mycorrhizal fungi induces symbiosis-specific *MtENOD11* expression in roots of *Medicago truncatula. Plant Physiol.* 131: 952-962.

Kosuta, S., Hazledine, S., Sun, J., Miwa, H., Morris, R.J., Downie, J.A. & Oldroyd, G.E.D. (2008) Differential and chaotic calcium signatures in the symbiosis signaling pathway of legumes. *Proc. Natl. Acad. Sci. USA* 105: 9823-9828.

Krajinski, F., Hause, B., Gianinazzi-Pearson, V. & Franken, P. (2002) *Mtha1*, a plasma membrane H^+-ATPase gene from *Medicago truncatula*, shows arbuscule-specific induced expression in mycorrhizal tissue. *Plant Biol.* 4: 754-761.

Kretzschmar, T., Kohlen, W., Sasse, J., Borghi, L., Schlegel, M., Bachelier, J.B., Reinhardt, D., Bours, R., Bouwmeester, H.J. & Martinoia, E. (2012) A petunia ABC protein controls strigolactone-dependent symbiotic signalling and branching. *Nature* 483: 341-344.

Krusell, L., Madsen, L.H., Sato, S., Aubert, G., Genua, A., Szczyglowski, K., Duc, G., Kaneko, T., Tabata, S., de Bruijn, F., Pajuelo, E., Sandal, N. & Stougaard, J. (2002) Shoot control of root development and nodulation is mediated by a receptor-like kinase. *Nature* 420: 422-426.

Krusell, L., Sato, N., Fukuhara, I., Koch, B.E.V., Grossmann, C., Okamoto, S., Oka-Kira, E., Otsubo, Y., Aubert, G., Nakagawa, T., Sato, S., Tabata, S., Duc, G., Parniske, M., Wang, T.L., Kawaguchi, M. & Stougaard, J. (2011) The *Clavata2* genes of pea and *Lotus japonicus* affect autoregulation of nodulation. *Plant J.* 65: 861-871.

Kubota, M., McGonigle, T.P. & Hyakumachi, M. (2005) Co-occurrence of *Arum-* and *Paris*-type morphologies of arbuscular mycorrhizae in cucumber and tomato. *Mycorrhiza* 15: 73-77.

Kuhn, H., Küster, H. & Requena, N. (2010) Membrane steroid-binding protein 1 induced by a diffusible fungal signal is critical for mycorrhization in *Medicago truncatula. New Phytol.* 185: 716-733.

Lammers, P.J., Jun, J., Abubaker, J., Arreola, R., Gopalan, A., Bago, B., Hernandez-Sebastia, C., Allen, J.W., Douds, D.D., Pfeffer, P.E. & Shachar-Hill, Y. (2001) The glyoxylate cycle in an arbuscular mycorrhizal fungus. Carbon flux and gene expression. *Plant Physiol.* 127: 1287-1298.

Landgraf, R., Schaarschmidt, S. & Hause, B. (2012) Repeated leaf wounding alters the colonization of *Medicago truncatula* roots by beneficial and pathogenic microorganisms. *Plant Cell Environ.* 35: 1344-1357.

Laurerssergues, D., Delaux, P.-M., Formey, D., Lelandais-Brière, C., Fort, S., Cottaz, S., Bécard, G., Niebel, A., Roux, C. & Combier, J.-P. (2012) The microRNA miR171h modulates arbuscular mycorrhizal colonization of *Medicago truncatula* by targeting *NSP2*. *Plant J.* 72: 512-522.

Lei, M. & Liu, D. (2011) Sucrose regulates plant responses to deficiencies in multiple nutrients. *Plant Signal. Behav.* 6: 1247-1249.

Lemoine, R., La Camera, S., Atanassova, R., Dédaldéchamp, F., Allario, T., Pourtau, N., Bonnemain, J.-L., Laloi, M., Coutos-Thévenot, P., Maurousset, L., Faucher, M., Girousse, C., Lemonnier, P., Parrilla, J. & Durand, M. (2013) Source to sink transport and regulation by environmental factors. *Front. Plant Sci.* 4: 272.

Lendenmann, M., Thonar, C., Barnard, R., Salmon, Y., Werner, R., Frossard, E. & Jansa, J. (2011) Symbiont identity matters: carbon and phosphorus fluxes between *Medicago truncatula* and different arbuscular mycorrhizal fungi. *Mycorrhiza* 21: 689-702.

Lendzemo, V., Kuyper, T.W. & Vierheilig, H. (2009) *Striga* seed-germination activity of root exudates and compounds present in stems of *Striga* host and nonhost (trap crop) plants is reduced due to root colonization by arbuscular mycorrhizal fungi. *Mycorrhiza* 19: 287-294.

Lerouge, P., Roche, P., Faucher, C., Maillet, F., Truchet, G., Prome, J.C. & Denarie, J. (1990) Symbiotic host-specificity of *Rhizobium meliloti* is determined by a sulphated and acylated glucosamine oligosaccharide signal. *Nature* 344: 781-784.

Lévy, J., Bres, C., Geurts, R., Chalhoub, B., Kulikova, O., Duc, G., Journet, E.-P., Ané, J.-M., Lauber, E., Bisseling, T., Dénarié, J., Rosenberg, C. & Debellé, F. (2004) A putative Ca^{2+} and calmodulin-dependent protein kinase required for bacterial and fungal symbiosis. *Science* 303: 1361-1364.

Li, G., Santoni, V. & Maurel, C. (2014) Plant aquaporins: roles in plant physiology. *Biochim. Biophys. Acta* 1840: 1574-1582.

Li, Y., Zou, Y.N., Huang, Y.M. & Wu, Q.S. (2012) Influence of arbuscular mycorrhizal fungi on sucrose metabolism of citrus seedlings. *Adv. Mat. Res.* 610-613: 3406-3409.

Libault, M., Joshi, T., Takahashi, K., Hurley-Sommer, A., Puricelli, K., Blake, S., Finger, R.E., Taylor, C.G., Xu, D., Nguyen, H.T. & Stacey, G. (2009) Large-scale analysis of putative soybean regulatory gene expression identifies a *Myb* gene involved in soybean nodule development. *Plant Physiol.* 151: 1207-1220.

Lim, C.W., Lee, Y.W. & Hwang, C.H. (2011) Soybean nodule-enhanced CLE peptides in roots act as signals in GmNARK-mediated nodulation suppression. *Plant Cell Physiol.* 52: 1613-1627.

Limpens, E., Franken, C., Smit, P., Willemse, J., Bisseling, T. & Greurts, R. (2003) LysM domain receptor kinases regulating rhizobial Nod factor-induced infection. *Science* 302: 630-633.

Lin, Y.-H., Ferguson, B.J., Kereszt, A. & Gresshoff, P.M. (2010) Suppression of hypernodulation in soybean by a leaf-extracted, NARK- and Nod factor-dependent, low molecular mass fraction. *New Phytol.* 185: 1074-1086.

Lioussanne, L., Jolicoeur, M. & St-Arnaud, M. (2009) Role of the modification in root exudation induced by arbuscular mycorrhizal colonization on the intraradical growth of *Phytophthora nicotianae* in tomato. *Mycorrhiza* 19: 443-448.

Liu, W., Kohlen, W., Lillo, A., Op den Camp, R., Ivanov, S., Hartog, M., Limpens, E., Jamil, M., Smaczniak, C., Kaufmann, K., Yang, W.-C., Hooiveld, G.J.E.J., Charnikhova, T., Bouwmeester, H.J., Bisseling, T. & Geurts, R. (2011) Strigolactone biosynthesis in *Medicago truncatula* and rice requires the symbiotic GRAS-type transcription factors NSP1 and NSP2. *Plant Cell* 23: 3853-3865.

Logi, C., Sbrana, C. & Giovannetti, M. (1998) Cellular events involved in survival of individual arbuscular mycorrhizal symbionts growing in the absence of the host. *Appl. Environ. Microbiol.* 64: 3473-3479.

Lohar, D.P. & Bird, D.M. (2003) *Lotus japonicus*: a new model to study root-parasitic nematodes. *Plant Cell Physiol.* 44: 1176-1184.

Lohse, S., Hause, B., Hause, G. & Fester, T. (2006) FtsZ characterization and immunolocalization in the two phases of plastid reorganization in arbuscular mycorrhizal roots of *Medicago truncatula. Plant Cell Physiol.* 47: 1124-1134.

Lohse, S., Schliemann, W., Ammer, C., Kopka, J., Strack, D. & Fester, T. (2005) Organization and metabolism of plastids and mitochondria in arbuscular mycorrhizal roots of *Medicago truncatula*. *Plant Physiol.* 139: 329-340.

López-Ráez, J.A., Charnikhova, T., Gómez-Roldán, V., Matusova, R., Kohlen, W., Vos, R.D., Verstappen, F., Puech-Pages, V., Bécard, G., Mulder, P. & Bouwmeester, H. (2008) Tomato strigolactones are derived from carotenoids and their biosynthesis is promoted by phosphate starvation. *New Phytol.* 178: 863-874.

López-Ráez, J.A., Kohlen, W., Charnikhova, T., Mulder, P., Undas, A.K., Sergeant, M.J., Verstappen, F., Bugg, T.D.H., Thompson, A.J., Ruyter-Spira, C. & Bouwmeester, H. (2010a) Does abscisic acid affect strigolactone biosynthesis? *New Phytol.* 187: 343-354.

López-Ráez, J.A., Verhage, A., Fernández, I., García, J., Azcón-Aguilar, C. & Pozo, M. (2010b) Hormonal and transcriptional profiles highlight common and differential host responses to arbuscular mycorrhizal fungi and the regulation of the oxylipin pathway. *J. Exp. Bot.* 61: 2589-2601.

Louarn, J., Carbonne, F., Delavault, P., Bécard, G. & Rochange, S. (2012) Reduced germination of *Orobanche cumana* seeds in the presence of arbuscular mycorrhizal fungi or their exudates. *PLoS ONE* 7: e49273.

Ludwig-Müller, J. (2010) Hormonal responses in host plants triggered by arbuscular mycorrhizal fungi. In: *Arbuscular Mycorrhizas: Physiology and Function*. H. Koltai & Y. Kapulnik (Eds.) Dordrecht: Springer. 169-190.

Ludwig-Müller, J. & Güther, M. (2007) Auxins as signals in arbuscular mycorrhiza formation. *Plant Signal. Behav.* 2: 194-196.

Madsen, E., Madsen, L., Radutoiu, S., Szczyglowski, K., Sato, S., Kaneko, T., Tabata, S., Sandal, N. & Stougaard, J. (2003) A receptor kinase gene of the LysM type is involved in legume perception of rhizobial signals. *Nature* 425: 637-640.

Maeda, D., Ashida, K., Iguchi, K., Chechetka, S.A., Hijikata, A., Okusako, Y., Deguchi, Y., Izui, K. & Hata, S. (2006) Knockdown of an arbuscular mycorrhiza-inducible phosphate transporter gene of *Lotus japonicus* suppresses mutualistic symbiosis. *Plant Cell Physiol.* 47: 807-817.

Magori, S. & Kawaguchi, M. (2010) Analysis of two potential long-distance signaling molecules, LjCLE-RS1/2 and jasmonic acid, in a hypernodulating mutant *too much love*. *Plant Signal Behav*. 5: 403-405.

Magori, S., Oka-Kira, E., Shibata, S., Umehara, Y., Kouchi, H., Hase, Y., Tanaka, A., Sato, S., Tabata, S. & Kawaguchi, M. (2009) *TOO MUCH LOVE*, a root regulator associated with the long-distance control of nodulation in *Lotus japonicus*. *Mol. Plant Microbe Interact.* 22: 259-268.

Maillet, F., Poinsot, V., Andre, O., Puech-Pages, V., Haouy, A., Gueunier, M., Cromer, L., Giraudet, D., Formey, D., Niebel, A., Martinez, E.A., Driguez, H., Bécard, G. & Denarie, J. (2011) Fungal lipochitooligosaccharide symbiotic signals in arbuscular mycorrhiza. *Nature* 469: 58-63.

Manthey, K., Krajinski, F., Hohnjec, N., Firnhaber, C., Pühler, A., Perlick, A.M. & Küster, H. (2004) Transcriptome profiling in root nodules and arbuscular mycorrhiza identifies a collection of novel genes induced during *Medicago truncatula* root endosymbioses. *Mol. Plant Microbe Interact.* 17: 1063-1077.

Markmann, K. & Parniske, M. (2009) Evolution of root endosymbiosis with bacteria: how novel are nodules? *Trends Plant Sci.* 14: 77-86.

Markmann, K., Giczey, G. & Parniske, M. (2008) Functional adaptation of a plant receptor-kinase paved the way for the evolution of intracellular root symbioses with bacteria. *PLoS Biol.* 6: e68.

Marsh, J.F., Rakocevic, A., Mitra, R.M., Brocard, L., Sun, J., Eschstruth, A., Long, S.R., Schultze, M., Ratet, P. & Oldroyd, G.E.D. (2007) *Medicago truncatula NIN* is essential for rhizobial-independent nodule organogenesis induced by autoactive calcium/calmodulin-dependent protein kinase. *Plant Physiol.* 144: 324-335.

Martín-Rodríguez, J.Á., León-Morcillo, R., Vierheilig, H., Ocampo, J.A., Ludwig-Müller, J. & García-Garrido, J.M. (2011) Ethylene-dependent/ethylene-independent ABA regulation of tomato plants colonized by arbuscular mycorrhiza fungi. *New Phytol.* 190: 193-205.

Martín Rodríguez, J.Á., León Morcillo, R., Vierheilig, H., Ocampo, J.A., Ludwig-Müller, J. & García Garrido, J.M. (2010) Mycorrhization of the notabilis and sitiens tomato mutants in relation to abscisic acid and ethylene contents. *J. Plant Physiol.* 167: 606-613.

Marx, C., Dexheimer, J., Gianinazzi-Pearson, V. & Gianinazzi, S. (1982) Enzymatic studies on the metabolism of vesicular-arbuscular mycorrhizas. IV. Ultracytoenzymological evidence (ATPase) for active transfer processes in the host-arbuscule interface. *New Phytol.* 90: 37-43.

McQueen-Mason, S., Durachko, D.M. & Cosgrove, D.J. (1992) Two endogenous proteins that induce cell wall extension in plants. *Plant Cell Envir.* 4: 1425-1433.

Meinild, A.-K., Klaerke, D.A. & Zeuthen, T. (1998) Bidirectional water fluxes and specificity for small hydrophilic molecules in aquaporins 0–5. *J. Biol. Chem* 273: 32446-32451.

Meixner, C., Ludwig-Mueller, J., Miersch, O., Gresshoff, P., Staehelin, C. & Vierheilig, H. (2005) Lack of mycorrhizal autoregulation and phytohormonal changes in the supernodulating soybean mutant *nts1007*. *Planta* 222: 709-715.

Meixner, C., Vegvarib, G., Ludwig-Müller, J., Gagnond, H., Steinkellner, S., Staehelin, C., Gresshoff, P. & Vierheilig, H. (2007) Two defined alleles of the LRR receptor kinase *GmNARK* in supernodulating soybean govern differing autoregulation of mycorrhization. *Physiol Plant.* 130: 261-270.

Mendgen, K. & Hahn, M. (2002) Plant infection and the establishment of fungal biotrophy. *Trends Plant Sci.* 7: 352-356.

Miller, J.B., Pratap, A., Miyahara, A., Zhou, L., Bornemann, S., Morris, R.J. & Oldroyd, G.E.D. (2013) Calcium/calmodulin-dependent protein kinase is negatively and positively regulated by calcium, providing a mechanism for decoding calcium responses during symbiosis signaling. *Plant Cell* 25: 5053-5066.

Mitra, R.M., Gleason, C.A., Edwards, A., Hadfield, J., Downie, J.A., Oldroyd, G.E.D. & Long, S.R. (2004) A Ca^{2+}/calmodulin-dependent protein kinase required for symbiotic nodule development: gene identification by transcript-based cloning. *Proc. Natl. Acad. Sci. USA* 101: 4701-4705.

Miya, A., Albert, P., Shinya, T., Desaki, Y., Ichimura, K., Shirasu, K., Narusaka, Y., Kawakami, N., Kaku, H. & Shibuya, N. (2007) CERK1, a LysM receptor kinase, is essential for chitin elicitor signaling in *Arabidopsis*. *Proc. Natl. Acad. Sci. USA* 104: 19613-19618.

Miyahara, A., Hirani, T.A., Oakes, M., Kereszt, A., Kobe, B., Djordjevic, M.A. & Gresshoff, P.M. (2008) Soybean nodule autoregulation receptor kinase phosphorylates two kinase-associated protein phosphatases *in vitro*. *J. Biol. Chem.* 283: 25381-25391.

Miyazawa, H., Oka-Kira, E., Sato, N., Takahashi, H., Wu, G.-J., Sato, S., Hayashi, M., Betsuyaku, S., Nakazono, M., Tabata, S., Harada, K., Sawa,

S., Fukuda, H. & Kawaguchi, M. (2010) The receptor-like kinase KLAVIER mediates systemic regulation of nodulation and non-symbiotic shoot development in *Lotus japonicus*. *Development* 137: 4317-4325.

Morandi, D., Branzanti, B. & Gianinazzi-Pearson, V. (1992) Effect of some plant flavonoids on in vitro behaviour of an arbuscular mycorrhizal fungus. *Agronomie 12*: 811-816.

Morandi, D., Sagan, M., Prado-Vivant, E. & Duc, G. (2000) Influence of genes determining supernodulation on root colonization by the mycorrhizal fungus *Glomus mosseae* in *Pisum sativum* and *Medicago truncatula* mutants. *Mycorrhiza* 10: 37-42.

Mortier, V., Den Herder, G., Whitford, R., Van de Velde, W., Rombauts, S., D'Haeseleer, K., Holsters, M. & Goormachtig, S. (2010) CLE peptides control *Medicago truncatula* nodulation locally and systemically. *Plant Physiol*. 153: 222-237.

Mukherjee, A. & Ané, J.-M. (2011) Germinating spore exudates from arbuscular mycorrhizal fungi: Molecular and developmental responses in plants and their regulation by ethylene. *Mol. Plant Microbe Interact*. 24: 260-270.

Murray, J.D., Muni, R.R.D., Torres-Jerez, I., Tang, Y., Allen, S., Andriankaja, M., Li, G., Laxmi, A., Cheng, X., Wen, J., Vaughan, D., Schultze, M., Sun, J., Charpentier, M., Oldroyd, G., Tadege, M., Ratet, P., Mysore, K.S., Chen, R. & Udvardi, M.K. (2011) *Vapyrin*, a gene essential for intracellular progression of arbuscular mycorrhizal symbiosis, is also essential for infection by rhizobia in the nodule symbiosis of *Medicago truncatula*. *Plant J*. 65: 244-252.

Nagahashi, G. & Douds, D.D. (1997) Appressorium formation by AM fungi on isolated cell walls of carrot roots. *New Phytol*. 136: 299-304.

Nagahashi, G. & Douds, D.D. (2000) Partial separation of root exudate components and their effects upon the growth of germinated spores of AM fungi. *Mycol. Res*. 104: 1453-1464.

Nagahashi, G. & Douds, D.D. (2007) Separated components of root exudate and cytosol stimulate different morphologically identifiable types of branching responses by arbuscular mycorrhizal fungi. *Mycol. Res*. 111: 487-492.

Nagahashi, G. & Douds, D.D. (2011) The effects of hydroxy fatty acids on the hyphal branching of germinated spores of AM fungi. *Fungal Biol*. 115: 351-358.

Nagy, R., Drissner, D., Amrhein, N., Jakobsen, I. & Bucher, M. (2009) Mycorrhizal phosphate uptake pathway in tomato is phosphorus-repressible and transcriptionally regulated. *New Phytol.* 181: 950-959.

Nagy, R., Karandashov, V., Chague, V., Kalinkevich, K., Tamasloukht, M.B., Xu, G., Jakobsen, I., Levy, A.A., Amrhein, N. & Bucher, M. (2005) The characterization of novel mycorrhiza-specific phosphate transporters from *Lycopersicon esculentum* and *Solanum tuberosum* uncovers functional redundancy in symbiotic phosphate transport in solanaceous species. *Plant J.* 42: 236-250.

Nagy, R., Vasconcelos, M.J.V., Zhao, S., McElver, J., Bruce, W., Amrhein, N., Raghothama, K.G. & Bucher, M. (2006) Differential regulation of five Pht1 phosphate transporters from maize (*Zea mays* L.). *Plant Biol.* 8: 186-197.

Nair, A. & Bhargava, S. (2012) Reduced mycorrhizal colonization (*rmc*) tomato mutant lacks expression of *SymRK* signaling pathway genes. *Plant Signal. Behav.* 7: 1578-1583.

Nair, M.G., Safir, G.R. & Siqueira, J.O. (1991) Isolation and identification of vesicular-arbuscular mycorrhiza-stimulatory compounds from clover (*Trifolium repens*) roots. *Appl. Environ. Microbiol.* 57: 434-439.

Nakagawa, T. & Kawaguchi, M. (2006) Shoot-applied MeJA suppresses root nodulation in *Lotus japonicus*. *Plant Cell Physiol.* 47: 176-180.

Nakagawa, T., Kaku, H., Shimoda, Y., Sugiyama, A., Shimamura, M., Takanashi, K., Yazaki, K., Aoki, T., Shibuya, N. & Kouchi, H. (2011) From defense to symbiosis: limited alterations in the kinase domain of LysM receptor-like kinases are crucial for evolution of legume–*Rhizobium* symbiosis. *Plant J.* 65: 169-180.

Navazio, L., Moscatiello, R., Genre, A., Novero, M., Baldan, B., Bonfante, P. & Mariani, P. (2007) A diffusible signal from arbuscular mycorrhizal fungi elicits a transient cytosolic calcium elevation in host plant cells. *Plant Physiol.* 144: 673-681.

Nehls, U. & Hampp, R. (2000) Carbon allocation in ectomycorrhizas. *Physiol. Mol. Plant Pathol.* 57: 95-100.

Nehls, U., Göhringer, F., Wittulsky, S. & Dietz, S. (2010) Fungal carbohydrate support in the ectomycorrhizal symbiosis: a review. *Plant Biol.* 12: 292-301.

Nishimura, R., Hayashi, M., Wu, G.-J., Kouchi, H., Imaizumi-Anraku, H., Murakami, Y., Kawasaki, S., Akao, S., Ohmori, M., Nagasawa, M., Harada, K. & Kawaguchi, M. (2002) HAR1 mediates systemic regulation of symbiotic organ development. *Nature* 420: 426-429.

Norman, J.R. & Hooker, J.E. (2000) Sporulation of *Phytophthora fragariae* shows greater stimulation by exudates of non-mycorrhizal than by mycorrhizal strawberry roots. *Mycol. Res.* 104: 1069-1073.

Nussaume, L., Kanno, S., Javot, H., Marin, E., Nakanishi, T.M. & Thibaud, M.-C. (2011) Phosphate import in plants: focus on the PHT1 transporters. *Front. Plant Sci.* 2: 83.

Nutman, P.S. (1952) Studies on the physiology of nodule formation: III. Experiments on the excision of root-tips and nodules. *Ann. Bot.* 16: 79-103.

Oelkers, K., Goffard, N., Weiller, G., Gresshoff, P., Mathesius, U. & Frickey, T. (2008) Bioinformatic analysis of the CLE signaling peptide family. *BMC Plant Biol.* 8: 1.

Oelkers, K., Goffard, N., Weiller, G., Gresshoff, P., Mathesius, U. & Frickey, T. (2009) Bioinformatic analysis of the CLE signaling peptide family. *BMC Plant Biol.* 9: 17.

Oka-Kira, E., Tateno, K., Miura, K.-i., Haga, T., Hayashi, M., Harada, K., Sato, S., Tabata, S., Shikazono, N., Tanaka, A., Watanabe, Y., Fukuhara, I., Nagata, T. & Kawaguchi, M. (2005) *klavier* (*klv*), A novel hypernodulation mutant of *Lotus japonicus* affected in vascular tissue organization and floral induction. *Plant J.* 44: 505-515.

Okamoto, S., Ohnishi, E., Sato, S., Takahashi, H., Nakazono, M., Tabata, S. & Kawaguchi, M. (2009) Nod factor/nitrate-induced *CLE* genes that drive HAR1-mediated systemic regulation of nodulation. *Plant Cell Physiol.* 50: 67-77.

Oláh, B., Brière, C., Bécard, G., Dénarié, J. & Gough, C. (2005) Nod factors and a diffusible factor from arbuscular mycorrhizal fungi stimulate lateral root formation in *Medicago truncatula* via the DMI1/DMI2 signalling pathway. *Plant J.* 44: 195-207.

Oldroyd, G.E.D. (2013) Speak, friend, and enter: signalling systems that promote beneficial symbiotic associations in plants. *Nat. Rev. Microbiol.* 11: 252-263.

Oldroyd, G.E.D. & Long, S.R. (2003) Identification and characterization of *Nodulation-Signaling Pathway 2*, a gene of *Medicago truncatula* involved in Nod factor signaling. *Plant Physiol.* 131: 1027-1032.

Olsson, O., Olsson, P.A. & Hammer, E.C. (2014) Phosphorus and carbon availability regulate structural composition and complexity of AM fungal mycelium. *Mycorrhiza* doi:10.1007/s00572-00014-00557-00578.

Op den Camp, R., Streng, A., De Mita, S., Cao, Q., Polone, E., Liu, W., Ammiraju, J.S.S., Kudrna, D., Wing, R., Untergasser, A., Bisseling, T. &

Geurts, R. (2011) LysM-type mycorrhizal receptor recruited for rhizobium symbiosis in nonlegume *Parasponia*. *Science* 331: 909-912.

Osipova, M.A., Mortier, V., Demchenko, K.N., Tsyganov, V.E., Tikhonovich, I.A., Lutova, L.A., Dolgikh, E.A. & Goormachtig, S. (2012) *WUSCHEL-RELATED HOMEOBOX5* gene expression and interaction of CLE peptides with components of the systemic control add two pieces to the puzzle of autoregulation of nodulation. *Plant Physiol.* 158: 1329-1341.

Paszkowski, U., Kroken, S., Roux, C. & Briggs, S.P. (2002) Rice phosphate transporters include an evolutionarily divergent gene specifically activated in arbuscular mycorrhizal symbiosis. *Proc. Natl. Acad. Sci. USA* 99: 13324-13329.

Pearson, J.N., Abbott, L.K. & Jasper, D.A. (1993) Mediation of competition between two colonizing VA mycorrhizal fungi by the host plant. *New Phytol.* 123: 93-98.

Pfeffer, P.E., Douds, D.D., Bécard, G. & Shachar-Hill, Y. (1999) Carbon uptake and the metabolism and transport of lipids in an arbuscular mycorrhiza. *Plant Physiol.* 120: 587-598.

Pingret, J.-L., Journet, E.-P. & Barker, D.G. (1998) *Rhizobium* Nod factor signaling: evidence for a G protein–mediated transduction mechanism. *Plant Cell* 10: 659-671.

Plet, J., Wasson, A., Ariel, F., Le Signor, C., Baker, D., Mathesius, U., Crespi, M. & Frugier, F. (2011) MtCRE1-dependent cytokinin signaling integrates bacterial and plant cues to coordinate symbiotic nodule organogenesis in *Medicago truncatula*. *Plant J.* 65: 622-633.

Poulin, M.J., Bel-Rhlid, R., Piché, Y. & Chênevert, R. (1993) Flavonoids released by carrot (*Daucus carota*) seedlings stimulate hyphal development of vesicular-arbuscular mycorrhizal fungi in the presence of optimal CO_2 enrichment. *J. Chem. Ecol.* 19: 2317-2327.

Proels, R.K. & Hückelhoven, R. (2014) Cell-wall invertases, key enzymes in the modulation of plant metabolism during defence responses. *Mol. Plant Pathol.* doi:10.1111/mpp.12139.

Pumplin, N. & Harrison, M.J. (2009) Live-cell imaging reveals periarbuscular membrane domains and organelle location in *Medicago truncatula* roots during arbuscular mycorrhizal symbiosis. *Plant Physiol.* 151: 809-819.

Pumplin, N., Mondo, S.J., Topp, S., Starker, C.G., Gantt, J.S. & Harrison, M.J. (2010) *Medicago truncatula* Vapyrin is a novel protein required for arbuscular mycorrhizal symbiosis. *Plant J.* 61: 482-494.

Radutoiu, S., Madsen, L., Madsen, E., Felle, H., Umehara, Y., Gronlund, M., Sato, S., Nakamura, Y. & Stougaard, J. (2003) Plant recognition of symbiotic bacteria requires two LysM receptor-like kinases. *Nature* 425: 585-592.

Rasmussen, A., Depuydt, S., Goormachtig, S. & Geelen, D. (2013) Strigolactones fine-tune the root system. *Planta* 238: 615-626.

Rausch, T. & Greiner, S. (2004) Plant protein inhibitors of invertases. *Biochim. Biophys. Acta* 1696: 253-261.

Rausch, C., Daram, P., Brunner, S., Jansa, J., Laloi, M., Leggewie, G., Amrhein, N. & Bucher, M. (2001) A phosphate transporter expressed in arbuscule-containing cells in potato. *Nature* 414: 462-466.

Ravnskov, S., Wu, Y. & Graham, J.H. (2003) Arbuscular mycorrhizal fungi differentially affect expression of genes coding for sucrose synthases in maize roots. *New Phytol.* 157: 539-545.

Reddy, D.M.R.S., Schorderet, M., Feller, U. & Reinhardt, D. (2007) A petunia mutant affected in intracellular accommodation and morphogenesis of arbuscular mycorrhizal fungi. *Plant J.* 51: 739-750.

Redecker, D., Kodner, R. & Graham, L.E. (2000) Glomalean fungi from the Ordovician. *Science* 15: 1920-1921.

Reid, D.E., Ferguson, B.J. & Gresshoff, P.M. (2011) Inoculation- and nitrate-induced CLE peptides of soybean control NARK-dependent nodule formation. *Mol. Plant Microbe Interact.* 24: 606-618.

Reid, D.E., Hayashi, S., Lorenc, M., Stiller, J., Edwards, D., Gresshoff, P.M. & Ferguson, B.J. (2012) Identification of systemic responses in soybean nodulation by xylem sap feeding and complete transcriptome sequencing reveal a novel component of the autoregulation pathway. *Plant Biotechnol. J.* 10: 680-689.

Remy, W., Taylor, T.N., Hass, H. & Kerp, H. (1994) Four hundred-million-year-old vesicular arbuscular mycorrhizae. *Proc. Natl. Acad. Sci. USA* 91: 11841-11843.

Requena, N., Breuninger, M., Franken, P. & Ocón, A. (2003) Symbiotic status, phosphate, and sucrose regulate the expression of two plasma membrane H^+-ATPase genes from the mycorrhizal fungus *Glomus mosseae. Plant Physiol.* 132: 1540-1549.

Riely, B.K., Lougnon, G., Ané, J.-M. & Cook, D.R. (2007) The symbiotic ion channel homolog DMI1 is localized in the nuclear membrane of *Medicago truncatula* roots. *Plant J.* 49: 208-216.

Rillig, M.C. (2004) Arbuscular mycorrhizae, glomalin, and soil aggregation. *Can. J. Soil Sci.* 84: 355-363.

Rivers, R.L., Dean, R.M., Chandy, G., Hall, J.E., Roberts, D.M. & Zeidel, M.L. (1997) Functional analysis of Nodulin 26, an aquaporin in soybean root nodule symbiosomes. *J. Biol. Chem.* 272: 16256-16261.

Rodiuc, N., Vieira, P., Banora, M.Y. & de Almeida Engler, J. (2014) On the track of transfer cell formation by specialized plant-parasitic nematodes. *Front. Plant Sci.* 5: 160.

Roitsch, T. & González, M.-C. (2004) Function and regulation of plant invertases: sweet sensations. *Trends Plant Sci.* 9: 606-613.

Roitsch, T., Balibrea, M.E., Hofmann, M., Proels, R. & Sinha, A.K. (2003) Extracellular invertase: key metabolic enzyme and PR protein. *J. Exp. Bot.* 54: 513-524.

Ruyter-Spira, C., Al-Babili, S., van der Krol, S. & Bouwmeester, H. (2013) The biology of strigolactones. *Trends Plant Sci.* 18: 72-83.

Sagan, M. & Duc, G. (1996) *Sym28* and *Sym29*, two new genes involved in regulation of nodulation in pea (*Pisum sativum* L). *Symbiosis* 20: 229-245.

Saito, M. (1995) Enzyme activities of the internal hyphae and germinated spores of an arbuscular mycorrhizal fungus, *Gigaspora margarita* Becker & Hall. *New Phytol.* 129: 425-431.

Saito, K., Yoshikawa, M., Yano, K., Miwa, H., Uchida, H., Asamizu, E., Sato, S., Tabata, S., Imaizumi-Anraku, H., Umehara, Y., Kouchi, H., Murooka, Y., Szczyglowski, K., Downie, J.A., Parniske, M., Hayashi, M. & Kawaguchi, M. (2007) NUCLEOPORIN85 is required for calcium spiking, fungal and bacterial symbioses, and seed production in *Lotus japonicus*. *Plant Cell* 19: 610-624.

Sakamoto, K. & Nohara, Y. (2009) Soybean (*Glycine max* [L.] Merr.) shoots systemically control arbuscule formation in mycorrhizal symbiosis. *Soil Sci. Plant Nutr.* 55: 252-257.

Scervino, J.M., Ponce, M.A., Erra-Bassells, R., Bompadre, J., Vierheilig, H., Ocampo, J.A. & Godeas, A.M. (2007) The effect of flavones and flavonols on colonization of tomato plants by arbuscular mycorrhizal fungi of the genera *Gigaspora* and *Glomus*. *Can. J. Microbiol.* 53: 702-709.

Scervino, J.M., Ponce, M.A., Erra-Bassells, R., Vierheilig, H., Ocampo, J.A. & Godeas, A.M. (2005a) Flavonoids exhibit fungal species and genus specific effects on the presymbiotic growth of *Gigaspora* and *Glomus*. *Mycol. Res.* 109: 789-794.

Scervino, J.M., Ponce, M.A., Erra-Bassells, R., Vierheilig, H., Ocampo, J.A. & Godeas, A.M. (2005b) Flavonoids exclusively present in mycorrhizal roots of white clover exhibit a different effect on arbuscular mycorrhizal fungi than flavonoids exclusively present in non-mycorrhizal roots of white clover. *J. Plant Interact.* 1: 15-22.

Schaarschmidt, S., Gonzalez, M.-C., Roitsch, T., Strack, D., Sonnewald, U. & Hause, B. (2007a) Regulation of arbuscular mycorrhization by carbon. The symbiotic interaction cannot be improved by increased carbon availability accomplished by root-specifically enhanced invertase activity. *Plant Physiol.* 143: 1827-1840.

Schaarschmidt, S., Gresshoff, P. & Hause, B. (2013) Analyzing the soybean transcriptome during autoregulation of mycorrhization identifies the transcription factors GmNF-YA1a/b as positive regulators of arbuscular mycorrhization. *Genome Biol.* 14: R62.

Schaarschmidt, S., Kopka, J., Ludwig-Müller, J. & Hause, B. (2007b) Regulation of arbuscular mycorrhization by apoplastic invertases: enhanced invertase activity in the leaf apoplast affects the symbiotic interaction. *Plant J.* 51: 390-405.

Schaarschmidt, S., Roitsch, T. & Hause, B. (2006) Arbuscular mycorrhiza induces gene expression of the apoplastic invertase LIN6 in tomato (*Lycopersicon esculentum*) roots. *J. Exp. Bot.* 57: 4015-4023.

Schaarschmidt, S., Qu, N., Strack, D., Sonnewald, U. & Hause, B. (2004) Local induction of the *alc* gene switch in transgenic tobacco plants by acetaldehyde. *Plant Cell Physiol.* 45: 1566-1577.

Schauser, L., Roussis, A., Stiller, J. & Stougaard, J. (1999) A plant regulator controlling development of symbiotic root nodules. *Nature* 402: 191-195.

Scheffknecht, S., Mammerler, R., Steinkellner, S. & Vierheilig, H. (2006) Root exudates of mycorrhizal tomato plants exhibit a different effect on microconidia germination of *Fusarium oxysporum* f. sp. *lycopersici* than root exudates from non-mycorrhizal tomato plants. *Mycorrhiza* 16: 365-370.

Schnabel, E., Journet, E.-P., Carvalho-Niebel, F., Duc, G. & Frugoli, J. (2005) The *Medicago truncatula SUNN* gene encodes a *CLV1*-like leucine-rich repeat receptor kinase that regulates nodule number and root length. *Plant Mol. Biol.* 58: 809-822.

Schliemann, W., Kolbe, B., Schmidt, J., Nimtz, M. & Wray, V. (2008) Accumulation of apocarotenoids in mycorrhizal roots of leek (*Allium porrum*). *Phytochemistry* 69: 1680-1688.

Schubert, A., Allara, P. & Morte, A. (2003) Cleavage of sucrose in roots of soybean (*Glycine max*) colonized by an arbuscular mycorrhizal fungus. *New Phytol.* 161: 495-501.

Schultz, C.J. & Harrison, M.J. (2008) Novel plant and fungal AGP-like proteins in the *Medicago truncatula–Glomus intraradices* arbuscular mycorrhizal symbiosis. *Mycorrhiza* 18: 403-412.

Schüßler, A. & Walker, C. (2010) *The Glomeromycota. A species List with new Families and new Genera.* Gloucester. Published in libraries at The Royal Botanic Garden Edinburgh, The Royal Botanic Garden Kew, Botanische Staatssammlung Munich and Oregon State University.

Schüßler, A. & Walker, C. (2011) Evolution of the 'plant-symbiotic' fungal phylum, Glomeromycota. In: *Evolution of Fungi and Fungal-Like Organisms*. S. Pöggeler & J. Wöstemeyer (Eds.) Berlin Heidelberg: *Springer*. 163-185.

Schüßler, A., Martin, H., Cohen, D., Fitz, M. & Wipf, D. (2006) Characterization of a carbohydrate transporter from symbiotic glomeromycotan fungi. *Nature* 444: 933-936.

Schüßler, A., Schwarzott, D. & Walker, C. (2001) A new fungal phylum, the *Glomeromycota*: phylogeny and evolution. *Mycol. Res.* 105: 1413-1421.

Searle, I.R., Men, A.E., Laniya, T.S., Buzas, D.M., Iturbe-Ormaetxe, I., Carroll, B.J. & Gresshoff, P.M. (2003) Long-distance signaling in nodulation directed by a CLAVATA1-like receptor kinase. *Science* 299: 109-112.

Seo, H.S., Li, J., Lee, S.-Y., Yu, J.-W., Kim, K.-H., Lee, S.-H., Lee, I.-J. & Paek, N.-C. (2007) The hypernodulating *nts* mutation induces jasmonate synthetic pathway in soybean leaves. *Mol. Cells* 24: 185-193.

Shachar-Hill, Y., Pfeffer, P.E., Douds, D.D., Osman, S.F., Doner, L.W. & Ratcliffe, R.G. (1995) Partitioning of intermediary carbon metabolism in vesicular-arbuscular mycorrhizal leek. *Plant Physiol.* 108: 7-15.

Shrihari, P.C., Sakamoto, K., Inubushi, K. & Akao, S. (2000) Interaction between supernodulating or non-nodulating mutants of soybean and two arbuscular mycorrhizal fungi. *Mycorrhiza* 10: 101-106.

Siciliano, V., Genre, A., Balestrini, R., Cappellazzo, G., deWit, P.J.G.M. & Bonfante, P. (2007) Transcriptome analysis of arbuscular mycorrhizal roots during development of the prepenetration apparatus. *Plant Physiol.* 144: 1455-1466.

Sieberer, B.J., Chabaud, M., Fournier, J., Timmers, A.C.J. & Barker, D.G. (2012) A switch in Ca^{2+} spiking signature is concomitant with endosymbiotic microbe entry into cortical root cells of *Medicago truncatula*. *Plant J.* 69: 822-830.

Sieberer, B.J., Chabaud, M., Timmers, A.C., Monin, A., Fournier, J. & Barker, D.G. (2009) A nuclear-targeted cameleon demonstrates intranuclear Ca^{2+}

spiking in *Medicago truncatula* root hairs in response to rhizobial nodulation factors. *Plant Physiol.* 151: 1197-1206.

Singh, S. & Parniske, M. (2012) Activation of calcium- and calmodulin-dependent protein kinase (CCaMK), the central regulator of plant root endosymbiosis. *Curr. Opin. Plant Biol.* 15: 444-453.

Singh, S., Katzer, K., Lambert, J., Cerri, M. & Parniske, M. (2014) CYCLOPS, a DNA-binding transcriptional activator, orchestrates symbiotic root nodule development. *Cell Host Micr.* 15: 139-152.

Sinha, A.K., Hofmann, M.G., Romer, U., Kockenberger, W., Elling, L. & Roitsch, T. (2002) Metabolizable and non-metabolizable sugars activate different signal transduction pathways in tomato. *Plant Physiol.* 128: 1480-1489.

Siqueira, J.O., Safir, G.R. & Nair, M.G. (1991) Stimulation of vesicular-arbuscular mycorrhiza formation and growth of white clover by flavonoid compounds. *New Phytol.* 118: 87-93.

Smit, P., Raedts, J., Portyanko, V., Debellé, F., Gough, C., Bisseling, T. & Geurts, R. (2005) NSP1 of the GRAS protein family is essential for rhizobial Nod-factor induced transcription. *Science* 308: 1789-1791.

Smith, F.A. & Smith, S.E. (1997) Structural diversity in (vesicular)-arbuscular mycorrhizal symbioses. *New Phytol.* 137: 373-388.

Smith, S.E., Dickson, S. & Smith, F.A. (2001) Nutrient transfer in arbuscular mycorrhizas: how are fungal and plant processes integrated? *Aust. J. Plant Pathol.* 28: 685-696.

Smith, S.E., Jakobsen, I., Grønlund, M. & Smith, F.A. (2011) Roles of arbuscular mycorrhizas in plant phosphorus nutrition: interactions between pathways of phosphorus uptake in arbuscular mycorrhizal roots have important implications for understanding and manipulating plant phosphorus acquisition. *Plant Physiol.* 156: 1050-1057.

Solaiman, M.Z. & Saito, M. (1997) Use of sugars by intraradical hyphae of arbuscular mycorrhizal fungi revealed by radiorespirometry. *New Phytol.* 136: 533-538.

Solaiman, M.Z., Senoo, K., Kawaguchi, M., Imaizumi-Anraku, H., Akao, S., Tanaka, A. & Obata, H. (2000) Characterization of mycorrhizas formed by *Glomus* sp. on roots of hypernodulating mutants of *Lotus japonicus*. *J. Plant Res.* 113: 443-448.

Sonnewald, U., Brauer, M., von Schaewen, A., Stitt, M. & Willmitzer, L. (1991) Transgenic tobacco plants expressing yeast-derived invertase in either the cytosol, vacuole or apoplast: a powerful tool for studying sucrose metabolism and sink/source interactions. *Plant J.* 1: 95-106.

Sonnewald, S., Priller, J.P.R., Schuster, J., Glickmann, E., Hajirezaei, M.-R., Siebig, S., Mudgett, M.B. & Sonnewald, U. (2012) Regulation of cell wall-bound invertase in pepper leaves by *Xanthomonas campestris* pv. *vesicatoria* type three effectors. *PLoS ONE* 7: e51763.

Soyano, T., Kouchi, H., Hirota, A. & Hayashi, M. (2013) NODULE INCEPTION directly targets *NF-Y* subunit genes to regulate essential processes of root nodule development in *Lotus japonicus*. *PLoS Genet.* 9: e1003352.

Staehelin, C., Charon, C., Boller, T., Crespi, M. & Kondorosi, Á. (2001) *Medicago truncatula* plants overexpressing the early nodulin gene *enod40* exhibit accelerated mycorrhizal colonization and enhanced formation of arbuscules. *Proc. Natl. Acad. Sci. USA* 98: 15366-15371.

Steinkellner, S., Lendzemo, V., Langer, I., Schweiger, P., Khaosaad, T., Toussaint, J.-P. & Vierheilig, H. (2007) Flavonoids and strigolactones in root exudates as signals in symbiotic and pathogenic plant-fungus interactions. *Molecules* 12: 1290-1306.

Stockinger, H., Walker, C. & Schüßler, A. (2009) '*Glomus intraradices* DAOM197198', a model fungus in arbuscular mycorrhiza research, is not *Glomus intraradices*. *New Phytol.* 183: 1176-1187.

Strack, D. & Fester, T. (2006) Isoprenoid metabolism and plastid reorganization in arbuscular mycorrhizal roots. *New Phytol.* 172: 22-34.

Stracke, S., Kistner, C., Yoshida, S., Mulder, L., Sato, S., Kaneko, T., Tabata, S., Sandal, N., Stougaard, J., Szczyglowski, K. & Parniske, M. (2002) A plant receptor-like kinase required for both bacterial and fungal symbiosis. *Nature* 417: 959-962.

Streng, A., op den Camp, R., Bisseling, T. & Geurts, R. (2011) Evolutionary origin of rhizobium Nod factor signaling. *Plant Signal. Behav.* 6: 1510-1514.

Strullu-Derrien, C. & Strullu, D.-G. (2007) Mycorrhization of fossil and living plants. *C. R. Palevol.* 6: 483-494.

Sturm, A. & Chrispeels, M.J. (1990) cDNA cloning of carrot extracellular beta-fructosidase and its expression in response to wounding and bacterial infection. *Plant Cell* 2: 1107-1119.

Sturm, A. & Tang, G.-Q. (1999) The sucrose-cleaving enzymes of plants are crucial for development, growth and carbon partitioning. *Trends Plant Sci.* 4: 401-407.

Sun, J., Cardoza, V., Mitchell, D.M., Bright, L., Oldroyd, G. & Harris, J.M. (2006) Crosstalk between jasmonic acid, ethylene and Nod factor signaling allows integration of diverse inputs for regulation of nodulation. *Plant J.* 46: 961-970.

Sun, S., Wang, J., Zhu, L., Liao, D., Gu, M., Ren, L., Kapulnik, Y. & Xu, G. (2012) An active factor from tomato root exudates plays an important role in efficient establishment of mycorrhizal symbiosis. *PLoS ONE* 7: e43385.

Swarbrick, P.J., Schulze-Lefert, P. & Scholes, J.D. (2006) Metabolic consequences of susceptibility and resistance (race-specific and broad-spectrum) in barley leaves challenged with powdery mildew. *Plant Cell Environ.* 29: 1061-1076.

Takeda, N., Sato, S., Asamizu, E., Tabata, S. & Parniske, M. (2009) Apoplastic plant subtilases support arbuscular mycorrhiza development in *Lotus japonicus*. *Plant J.* 58: 766-777.

Talbot, N.J. (2010) Living the sweet life: how does a plant pathogenic fungus acquire sugar from plants? *PLoS Biol.* 8: e1000308.

Talukdar, T., Gorecka, K., de Carvalho-Niebel, F., Downie, J., Cullimore, J. & Pikula, S. (2009) Annexins – calcium- and membrane-binding proteins in the plant kingdom: potential role in nodulation and mycorrhization in *Medicago truncatula*. *Acta Biochim. Pol.* 56: 199-210.

Tamasloukht, M.B., Séjalon-Delmas, N., Kluever, A., Jauneau, A., Roux, C., Bécard, G. & Franken, P. (2003) Root factors induce mitochondrial-related gene expression and fungal respiration during the developmental switch from asymbiosis to presymbiosis in the arbuscular mycorrhizal fungus *Gigaspora rosea*. *Plant Physiol.* 131: 1468-1478.

Tambalo, D.D., Vanderlinde, E.M., Robinson, S., Halmillawewa, A., Hynes, M.F. & Yost, C.K. (2014) Legume seed exudates and *Physcomitrella patens* extracts influence swarming behavior in *Rhizobium leguminosarum*. *Can. J. Microbiol.* 60: 15-24.

Tamura, Y., Kobae, Y., Mizuno, T. & Hata, S. (2012) Identification and expression analysis of arbuscular mycorrhiza-inducible phosphate transporter genes of soybean. *Biosci. Biotechnol. Biochem.* 76: 309-313.

Tan, Z., Hu, Y. & Lin, Z. (2012) PhPT4 is a mycorrhizal-phosphate transporter suppressed by lysophosphatidylcholine in petunia roots. *Plant Mol. Biol. Rep.* 30: 1480-1487.

Tejeda-Sartorius, M., Vega, O.M.d.l. & Délano-Frier, J.P. (2008) Jasmonic acid influences mycorrhizal colonization in tomato plants by modifying the expression of genes involved in carbohydrate partitioning. *Physiol. Plant.* 133: 339-353.

Tester, M., Smith, F.A. & Smith, S.E. (1985) Phosphate inflow into *Trifolium subterraneum* L.: effects of photon irradiance and mycorrhizal infection. *Soil Biol. Biochem.* 17: 807-810.

Thoma, I., Loeffler, C., Sinha, A.K., Gupta, M., Krischke, M., Steffan, B., Roitsch, T. & Mueller, M.J. (2003) Cyclopentenone isoprostanes induced by reactive oxygen species trigger defense gene activation and phytoalexin accumulation in plants. *Plant J.* 34: 363-375.

Tisserant, E., Kohler, A., Dozolme-Seddas, P., Balestrini, R., Benabdellah, K., Colard, A., Croll, D., Da Silva, C., Gomez, S.K., Koul, R., Ferrol, N., Fiorilli, V., Formey, D., Franken, P., Helber, N., Hijri, M., Lanfranco, L., Lindquist, E., Liu, Y., Malbreil, M., Morin, E., Poulain, J., Shapiro, H., van Tuinen, D., Waschke, A., Azcón-Aguilar, C., Bécard, G., Bonfante, P., Harrison, M.J., Küster, H., Lammers, P., Paszkowski, U., Requena, N., Rensing, S.A., Roux, C., Sanders, I.R., Shachar-Hill, Y., Tuskan, G., Young, J.P.W., Gianinazzi-Pearson, V. & Martin, F. (2012) The transcriptome of the arbuscular mycorrhizal fungus *Glomus intraradices* (DAOM 197198) reveals functional tradeoffs in an obligate symbiont. *New Phytol.* 193: 755-769.

Tisserant, E., Malbreil, M., Kuo, A., Kohler, A., Symeonidi, A., Balestrini, R., Charron, P., Duensing, N., Frei dit Frey, N., Gianinazzi-Pearson, V., Gilbert, L.B., Handa, Y., Herr, J.R., Hijri, M., Koul, R., Kawaguchi, M., Krajinski, F., Lammers, P.J., Masclaux, F.G., Murat, C., Morin, E., Ndikumana, S., Pagni, M., Petitpierre, D., Requena, N., Rosikiewicz, P., Riley, R., Saito, K., San Clemente, H., Shapiro, H., van Tuinen, D., Bécard, G., Bonfante, P., Paszkowski, U., Shachar-Hill, Y.Y., Tuskan, G.A., Young, P.W., Sanders, I.R., Henrissat, B., Rensing, S.A., Grigoriev, I.V., Corradi, N., Roux, C. & Martin, F. (2013) Genome of an arbuscular mycorrhizal fungus provides insight into the oldest plant symbiosis. *Proc. Natl. Acad. Sci. USA* 110: 20117-20122.

Tirichine, L., Imaizumi-Anraku, H., Yoshida, S., Murakami, Y., Madsen, L.H., Miwa, H., Nakagawa, T., Sandal, N., Albrektsen, A.S., Kawaguchi, M., Downie, A., Sato, S., Tabata, S., Kouchi, H., Parniske, M., Kawasaki, S. & Stougaard, J. (2006) Deregulation of a Ca2+/calmodulin-dependent kinase leads to spontaneous nodule development. *Nature* 441: 1153-1156.

Tsai, S.M. & Phillips, D.A. (1991) Flavonoids released naturally from alfalfa promote development of symbiotic *Glomus* spores in vitro. *Appl. Environ. Microbiol.* 57: 1485-1488.

Tymowska-Lalanne, Z. & Kreis, M. (1998) The plant invertases: physiology, biochemistry and molecular biology. *Adv. Bot. Res.* 28: 71-117.

Uehlein, N., Fileschi, K., Eckert, M., Bienert, G.P., Bertl, A. & Kaldenhoff, R. (2007) Arbuscular mycorrhizal symbiosis and plant aquaporin expression. *Phytochemistry* 68: 122-129.

van Aarle, I.M. & Olsson, P.A. (2003) Fungal lipid accumulation and development of mycelial structures by two arbuscular mycorrhizal fungi. *Appl. Environ. Microbiol.* 69: 6762-6767.

van der Heijden, M.G.A. & Horton, T.R. (2009) Socialism in soil? The importance of mycorrhizal fungal networks for facilitation in natural ecosystems. *J. Ecol.* 97: 1139-1150.

van der Heijden, M.G.A., Klironomos, J.N., Ursic, M., Moutoglis, P., Streitwolf-Engel, R., Boller, T., Wiemken, A. & Sanders, I.R. (1998) Mycorrhizal fungal diversity determines plant biodiversity, ecosystem variability and productivity. *Nature* 396: 69-72.

vanKuyk, P.A., Diderich, J.A., MacCabe, A.P., Hererro, O., Ruijter, G.J.G. & Visser, J. (2004) *Aspergillus niger mstA* encodes a high-affinity sugar/H^+ symporter which is regulated in response to extracellular pH. *Biochem J.* 379: 375-383.

van Noorden, G.E., Kerim, T., Goffard, N., Wiblin, R., Pellerone, F.I., Rolfe, B.G. & Mathesius, U. (2007) Overlap of proteome changes in *Medicago truncatula* in response to auxin and *Sinorhizobium meliloti*. *Plant Physiol.* 144: 1115-1131.

van Noorden, G.E., Ross, J.J., Reid, J.B., Rolfe, B.G. & Mathesius, U. (2006) Defective long-distance auxin transport regulation in the *Medicago truncatula* super numeric nodules mutant. *Plant Physiol.* 140: 1494-1506.

Vierheilig, H. (2004) Further root colonization by arbuscular mycorrhizal fungi in already mycorrhizal plants is suppressed after a critical level of root colonization. *J. Plant Physiol.* 161: 339-341.

Vierheilig, H., Coughlan, A.P., Wyss, U. & Piché, Y. (1998) Ink and vinegar, a simple staining technique for arbuscular-mycorrhizal fungi. *Appl. Environ. Microbiol.* 64: 5004-5007.Vierheilig, H., Gagnon, H., Strack, D. & Maier, W. (2000a) Accumulation of cyclohexenone derivatives in barley, wheat and maize roots in response to inoculation with different arbuscular mycorrhizal fungi. *Mycorrhiza* 9: 291-293.

Vierheilig, H., Garcia-Garrido, J.M., Wyss, U. & Piché, Y. (2000b) Systemic suppression of mycorrhizal colonization of barley roots already colonized by AM fungi. *Soil Biol. Biochem.* 32: 589-595.

Vierheilig, H., Lerat, S. & Piché, Y. (2003) Systemic inhibition of arbuscular mycorrhiza development by root exudates of cucumber plants colonized by *Glomus mosseae*. *Mycorrhiza* 13: 167-170.

Vierheilig, H., Maier, W., Wyss, U., Samson, J., Strack, D. & Piché, Y. (2000c) Cyclohexanone derivative- and phosphate-levels in split-root systems and their role in the systemic suppression of mycorrhization in precolonized barley plants. *J. Plant Physiol.* 157: 593-599.

von Schaewen, A., Stitt, M., Schmidt, R., Sonnewald, U. & Willmitzer, L. (1990) Expression of a yeast-derived invertase in the cell wall of tobacco and *Arabidopsis* plants leads to accumulation of carbohydrate and inhibition of photosynthesis and strongly influences growth and phenotype of transgenic tobacco plants. *EMBO J.* 9: 3033-3044.

Wagg, C., Bender, S.F., Widmer, F. & van der Heijden, M.G.A. (2014) Soil biodiversity and soil community composition determine ecosystem multifunctionality. *Proc. Natl. Acad. Sci. USA* 111: 5266-5270.

Wall, L.G. (2000) The actinorhizal symbiosis. *J. Plant Growth Regul.* 19: 167-182.

Wallace, I.S., Choi, W.-G. & Roberts, D.M. (2006) The structure, function and regulation of the nodulin 26-like intrinsic protein family of plant aquaglyceroporins. *Biochim. Biophys. Acta* 1758: 1165-1175.

Wallace, I.S., Wills, D.M., Guenther, J.F. & Roberts, D.M. (2002) Functional selectivity for glycerol of the nodulin 26 subfamily of plant membrane intrinsic proteins. *FEBS Lett.* 523: 109-112.

Walter, M.H., Flo, D.S., Hans, J., Fester, T. & Strack, D. (2007) Apocarotenoid biosynthesis in arbuscular mycorrhizal roots: Contributions from methylerythritol phosphate pathway isogenes and tools for its manipulation. *Phytochemistry* 68: 130-138.

Wang, E., Schornack, S., Marsh, John F., Gobbato, E., Schwessinger, B., Eastmond, P., Schultze, M., Kamoun, S. & Oldroyd, Giles E.D. (2012) A common signaling process that promotes mycorrhizal and oomycete colonization of plants. *Curr. Biol.* 22: 2242-2246.

Wang, G. & Fiers, M. (2010) CLE peptide signaling during plant development. *Protoplasma* 240: 33-43.

Waters, M.T., Nelson, D.C., Scaffidi, A., Flematti, G.R., Sun, Y.K., Dixon, K.W. & Smith, S.M. (2012) Specialisation within the DWARF14 protein family confers distinct responses to karrikins and strigolactones in *Arabidopsis*. *Development* 139: 1285-1295.

Weerasinghe, R.R., Bird, D.M. & Allen, N.S. (2005) Root-knot nematodes and bacterial Nod factors elicit common signal transduction events in *Lotus japonicus*. *Proc. Natl. Acad. Sci. USA* 102: 3147-3152.

Weidmann, S., Sanchez, L., Descombin, J., Chatagnier, O., Gianinazzi, S. & Gianinazzi-Pearson, V. (2004) Fungal elicitation of signal transduction-related plant genes precedes mycorrhiza establishment and requires the *dmi3* gene in *Medicago truncatula*. *Mol. Plant Microbe Interact.* 17: 1385-1393.

Weig, A.R. & Jakob, C. (2000) Functional identification of the glycerol permease activity of *Arabidopsis thaliana* NLM1 and NLM2 proteins by heterologous expression in *Saccharomyces cerevisiae*. *FEBS Lett.* 481: 293-298.

Willmann, R., Lajunen, H.M., Erbs, G., Newman, M.-A., Kolb, D., Tsuda, K., Katagiri, F., Fliegmann, J., Bono, J.-J., Cullimore, J.V., Jehle, A.K., Götz, F., Kulik, A., Molinaro, A., Lipka, V., Gust, A.A. & Nürnberger, T. (2011) *Arabidopsis* lysin-motif proteins LYM1 LYM3 CERK1 mediate bacterial peptidoglycan sensing and immunity to bacterial infection. *Proc. Natl. Acad. Sci. USA* 108: 19824-19829.

Wopereis, J., Pajuelo, E., Dazzo, F.B., Jiang, Q., Gresshoff, P.M., Bruijn, F.J.d., Stougaard, J. & Szczyglowski, K. (2000) Short root mutant of *Lotus japonicus* with a dramatically altered symbiotic phenotype. *Plant J.* 23: 97-114.

Wright, D.P., Read, D.J. & Scholes, J.D. (1998) Mycorrhizal sink strength influences whole plant carbon balance of *Trifolium repens* L. *Plant Cell Environ.* 21: 881-891.

Yang, S.-Y. & Paszkowski, U. (2011) Phosphate import at the arbuscule: just a nutrient? *Mol. Plant Microbe Interact.* 24: 1296-1299.

Yano, K., Yoshida, S., Müller, J., Singh, S., Banba, M., Vickers, K., Markmann, K., White, C., Schuller, B., Sato, S., Asamizu, E., Tabata, S., Murooka, Y., Perry, J., Wang, T.L., Kawaguchi, M., Imaizumi-Anraku, H., Hayashi, M. & Parniske, M. (2008) CYCLOPS, a mediator of symbiotic intracellular accommodation. *Proc. Natl. Acad. Sci. USA* 105: 20540-20545.

Yoneyama, K., Xie, X., Kusumoto, D., Sekimoto, H., Sugimoto, Y., Takeuchi, Y. & Yoneyama, K. (2007a) Nitrogen deficiency as well as phosphorus deficiency in sorghum promotes the production and exudation of 5-deoxystrigol, the host recognition signal for arbuscular mycorrhizal fungi and root parasites. *Planta* 227: 125-132.

Yoneyama, K., Xie, X., Sekimoto, H., Takeuchi, Y., Ogasawara, S., Akiyama, K., Hayashi, H. & Yoneyama, K. (2008) Strigolactones, host recognition signals for root parasitic plants and arbuscular mycorrhizal fungi, from Fabaceae plants. *New Phytol.* 179: 484-494.

Yoneyama, K., Yoneyama, K., Takeuchi, Y. & Sekimoto, H. (2007b) Phosphorus deficiency in red clover promotes exudation of orobanchol, the signal for mycorrhizal symbionts and germination stimulant for root parasites. *Planta* 225: 1031-1038.

Yoshida, S., Kameoka, H., Tempo, M., Akiyama, K., Umehara, M., Yamaguchi, S., Hayashi, H., Kyozuka, J. & Shirasu, K. (2012) The D3 F-box protein is a key component in host strigolactone responses essential for arbuscular mycorrhizal symbiosis. *New Phytol.* 196: 1208-1216.

Young, N.D., Debelle, F., Oldroyd, G.E.D., Geurts, R., Cannon, S.B., Udvardi, M.K., Benedito, V.A., Mayer, K.F.X., Gouzy, J., Schoof, H., Van de Peer, Y., Proost, S., Cook, D.R., Meyers, B.C., Spannagl, M., Cheung, F., De Mita, S., Krishnakumar, V., Gundlach, H., Zhou, S., Mudge, J., Bharti, A.K., Murray, J.D., Naoumkina, M.A., Rosen, B., Silverstein, K.A.T., Tang, H., Rombauts, S., Zhao, P.X., Zhou, P., Barbe, V., Bardou, P., Bechner, M., Bellec, A., Berger, A., Berges, H., Bidwell, S., Bisseling, T., Choisne, N., Couloux, A., Denny, R., Deshpande, S., Dai, X., Doyle, J.J., Dudez, A.-M., Farmer, A.D., Fouteau, S., Franken, C., Gibelin, C., Gish, J., Goldstein, S., Gonzalez, A.J., Green, P.J., Hallab, A., Hartog, M., Hua, A., Humphray, S.J., Jeong, D.-H., Jing, Y., Jocker, A., Kenton, S.M., Kim, D.-J., Klee, K., Lai, H., Lang, C., Lin, S., Macmil, S.L., Magdelenat, G., Matthews, L., McCorrison, J., Monaghan, E.L., Mun, J.-H., Najar, F.Z., Nicholson, C., Noirot, C., O/'Bleness, M., Paule, C.R., Poulain, J., Prion, F., Qin, B., Qu, C., Retzel, E.F., Riddle, C., Sallet, E., Samain, S., Samson, N., Sanders, I., Saurat, O., Scarpelli, C., Schiex, T., Segurens, B., Severin, A.J., Sherrier, D.J., Shi, R., Sims, S., Singer, S.R., Sinharoy, S., Sterck, L., Viollet, A., Wang, B.-B., Wang, K., Wang, M., Wang, X., Warfsmann, J., Weissenbach, J., White, D.D., White, J.D., Wiley, G.B., Wincker, P., Xing, Y., Yang, L., Yao, Z., Ying, F., Zhai, J., Zhou, L., Zuber, A., Denarie, J., Dixon, R.A., May, G.D., Schwartz, D.C., Rogers, J., Quetier, F., Town, C.D. & Roe, B.A. (2011) The *Medicago* genome provides insight into the evolution of rhizobial symbioses. *Nature* 480: 520-524.

Yu, Q., Tang, C. & Kuo, J. (2000) A critical review on methods to measure apoplastic pH in plants. *Plant Soil* 219: 29-40.

Zardoya, R., Ding, X., Kitagawa, Y. & Chrispeels, M.J. (2002) Origin of plant glycerol transporters by horizontal gene transfer and functional recruitment. *Proc. Natl. Acad. Sci. USA* 99: 14893-14896.

Zhang, L., Cohn, N.S. & Mitchell, J.P. (1996) Induction of a pea cell-wall invertase gene by wounding and its localized expression in phloem. *Plant Physiol.* 112: 1111-1117.

Zhang, Q., Blaylock, L.A. & Harrison, M.J. (2010) Two *Medicago truncatula* half-ABC transporters are essential for arbuscule development in arbuscular mycorrhizal symbiosis. *Plant Cell* 22: 1483-1497.

Zhu, H., Riely, B.K., Burns, N.J. & Ané, J.-M. (2006) Tracing nonlegume orthologs of legume genes required for nodulation and arbuscular mycorrhizal symbioses. *Genetics* 172: 2491-2499.

Zwanenburg, B. & Pospíšil, T. (2013) Structure and activity of strigolactones: new plant hormones with a rich future. *Mol. Plant* 6: 38-62.

In: Mycorrhizas
Editor: Elaine Warwick

ISBN: 978-1-63321-436-1

Chapter 2

ARBUSCULAR MYCORRHIZAL STRUCTURES IN NATURAL AND DISTURBED SITES

Marcela Claudia Pagano*
Department of Physics, Federal University of Minas Gerais, Belo Horizonte, Minas Gerais, Brazil

ABSTRACT

Most plant species associate with mycorrhizas to perform better under stressful conditions and the increasing recognition of the mycorrhizal interactions, especially arbuscular mycorrhizal fungi (AMF) may have important functions in terrestrial ecosystems. AMF structures such as total root colonized and percent root length colonized by arbuscules can be used to estimate of AMF abundance within a plant or ecosystem. Also important is to evaluate the extraradical hyphae of AMF that extend into the surrounding soil much further than do roots. Due to their strategic position at the soil–root interface, the study of AMF in different undisturbed or degraded ecosystems can contribute to better understand the effect of global change on plant-soil associated communities. Many plant species in tropical regions where semiarid environments are also found associate with AMF and this symbioses has to be monitored under global change. This chapter explores the current available information relevant to benefits from AMF symbioses in natural

* Corresponding author: Dr Marcela Claudia Pagano, Federal University of Minas Gerais, Av. Antônio Carlos 6627, Belo Horizonte, Brazil. Email: marpagano@gmail.com; mcp2006 @ufmg.br.

and disturbed sites. Possible functions of AMF structures and research on Glomalin and AMF interaction with biochar are also discussed. Better knowledge of the wide variation in AMF parameters is important for understanding the ecology of different land ecosystems and for management purposes.

Keywords: Fungal structures, mycorrhizas, glomalin, tropical biomes, semiarid

INTRODUCTION

Interest in ecosystem services (Redondo-Brenes and Montagnini 2010, Sala et al. 2000, van der Wal et. al. 2012) is increasingly appreciated. The ecosystem services of soils and the impact of agriculture on soil structure or changing soil species inhabitants have been slowly recognized (Wall and Nielsen 2012, Pagano 2013). In particular, soils and its biota are essential for agricultural production and thus, provision of a high proportion of our food (Wall and Nielsen 2012). Due to its high biodiversity, it was proposed that most species of soil biota are redundant and will not affect the rate of an ecosystem function or the services provided. This species richness can led to greater functioning, but the presence of key species in soil communities can allow to disproportionately large effect on function compared to the number of species (reviewed by Nielsen et al. 2011).

Most plant species associate with mycorrhizas to perform better under stressful conditions and the increasing recognition of the mycorrhizal interactions, especially arbuscular mycorrhizal fungi (AMF) may have important functions in terrestrial ecosystems (Pagano 2012, Simard and Austin 2010). AMF structures such as total root colonized and percent root length colonized by arbuscules can be used to estimate of AMF abundance within a plant or ecosystem (Treseder 2013). The extraradical hyphae of AMF that extend into the surrounding soil much further than do roots was few studied.

As microorganism present both at the soil and root, the study of AMF in different ecosystems can contribute to better understand the effect of global change on plant-soil associated communities (Pagano 2013, Pagano and Dhar 2014). Many plant species in tropical regions where semiarid environments are also found associate with AMF and this symbioses has to be monitored under global change.

This chapter explores the current available information relevant to benefits from AMF symbioses in natural and disturbed sites. Possible functions of AMF structures and research on Glomalin and AMF interaction with biochar are also discussed. Better knowledge of the wide variation in AMF parameters is important for understanding the ecology of different land ecosystems and for management purposes. This review will consider the range of AMF structures (infection, morphology, structure) and show how many of these are confined to one or other hemisphere. How occurrence of the different genera and species of AMF might be related to soil conditions will also be discussed. Better knowledge of the wide variation in mycorrhizal parameters is important for understanding the ecology of different regions and for management purposes. Morphological characters are of great potential use for taxonomy and also as indicators of land management.

Ecosystem Services of Mycorrhizas

Interest in ecosystem services and the recognition that global and climatic changes can negatively affect biodiversity and ecosystem function (Pagano 2013, Redondo-Brenes and Montagnini 2010, Sala et al. 2000, van der Wal et. al. 2012, is increasingly appreciated. Likewise, interest in rainforest protection and ecologically sustainable land-use such as the agroforestry systems is receiving more interest due to the adverse climate change potential impacts on agricultural intensification (Borie et al. 2006, Duarte et al. 2013), and increased carbon (C) sequestration in terrestrial ecosystems is required for climate change mitigation (Pagano and Dhar 2014). To adapt to the changing climate and to meet the needs of growing population for food, fodder and fuel, the implementation of better soil management practices is necessary. Soil restoration and enhancement of soil fertility are also necessary (Reviewed by Lal 2009). Sustainable soil management system in the tropics include seed bed preparation, crop rotations, fertility management, etc. (See Table 1 in Lal 2009).

It is known that the soil C sequestration is the largest of terrestrial organic C stocks also being a major source of atmospheric carbon dioxide (CO_2). Rising atmospheric CO_2 seems to stimulate plant growth and soil C input but may in addition alter microbial decomposition. However, the combined effect of these responses on long-term C storage is uncertain and participate in climate change (Covacevich et al. 2012, van Groenigen et al. 2014).

The actual soil degradation, given by negligence, has lead to erosion, salinization and nutrient depletion. The decrease in SOM reserves and reduced soil structure have resulted in the proposed technology of no-tillage; residue mulch and cover crops; integrated nutrient management; biochar addition and improved genetically modified crops (Reviewed by Lal 2009). It is known that erosion reduces seed germination, stand establishment, plant growth and vigor. Moreover, it increases risks of inundation, pesticide damage, and seedling burial. However, the many experimental data, e.g. of tropical crops impact due to erosion, are scarce. This has resulted in the use of fertilizers and soil amendments (Reviewed by Lal 2009).

The valorization of the ecological services of riparian forests and wetlands for a healthy hydrological cycle (e.g. riparian forests protect the water quality of rivers, lakes and reservoirs and wetlands control floods, sedimentation and regulate the water quality) is fortunately increasing (Tundisi and Tundisi 2010). The preservation of certain plant species such as *Typha* spp., which occur in wetlands, was indicated due to their importance. Additionally, some *Typha* species associate with AMF (Wang and Qiu 2006).

The ecosystem services of soils are often not recognized and the impact of agriculture on soil structure (essential for facilitating water infiltration, success of sustainable agriculture, and preventing erosion and flooding) or changing soil species inhabitants need urgent studies. Moreover, it is known that restored forests can improve ecosystem services and enhance biodiversity conservation; however, to arrive at the same composition and structure of the original forest cover is not possible. Thus, tactics to restoring forest ecosystems depend on levels of soil degradation and desired restoration results (Chazdon 2008). Consequently, efforts to restore forests, ecosystem services, and biodiversity worldwide must combine with sustainable rural incomes and community participation which require adaptive management as dynamic, resilient systems that can resist anthropogenic stresses e.g. climate change and habitat fragmentation (Chazdon 2008).

Ecosystem services depend on soil properties, soil conditions (e.g. moisture, temperature) and on the biological processes (species interactions e.g. symbiosis, competition) as well as on the ecosystem management (e.g. tillage), which will select organisms that occur in the ecosystem (Dominati et al. 2010, Pagano 2011). Approaches on recovery of ecosystem services by drawing on soil services, soil biodiversity and on current thinking on ecosystem restoration to attain a better understanding of ecosystems need connectivity areas such as corridors assisting species to respond and survive the biome shifts caused by global change. However, it will involve regional understanding and cooperation between researchers, policy makers and the community. It is known that number, diversity, and activity of both free and symbiotic fungal populations are modified by plantations and crops (Cardoso and Kuyper 2006, Kahiluoto et al. 2009, Nyfeler et al. 2011, Pagano et al. 2011).

Among ecosystem services, the AMF protect soil structure and tree roots against disease or drought (Simard and Austin 2010). Moreover, mycorrhizas can contribute to sustainable plant production, ecosystem restoration, and soil C sequestration mitigating global climate change (Warnock et al. 2007) (See below). AMF constitute an important component of the soil microbial community and are very successful microfungi that form mutualistic symbioses with 80% of all plant species. AMF participate in ecosystem services such as increasing soil structure, protecting soil carbon (C) against mineralization, and protecting tree roots against disease or drought (Simard and Austin 2010). Moreover, it is known that mycorrhizas and black C (biochar) are theoretically essential in various soil ecosystem services, contributing to sustainable plant production, ecosystem restoration, and soil C sequestration and hence mitigation of global climate change (Warnock et al. 2007).

Most plant species associate with mycorrhizas to perform better under stressful conditions and the increasing recognition of the mycorrhizal interactions, especially by AMF may have important functions in terrestrial ecosystems. AMF structures such as total root colonized, percent root length colonized by arbuscules can be used to estimate of AMF abundance within a plant or ecosystem. The extraradical hyphae of AMF extend into the surrounding soil much further than do roots. Among AMF structures, arbuscules are regularly the primary site of nutrient and C exchange between the fungus and plant (Smith and Read 2008). However, the role of arbuscules has not been fully confirmed and other exchange sites—including coiling AMF—are probable sites (Treseder 2013).

Due to their strategical position at the soil–root interface, the study of AMF in different undisturbed or degraded ecosystems can contribute to better understand the effect of global change on plant-soil associated communities. Many plant species in tropical regions where semiarid environments are also found associate with AMF and this symbioses has to be monitored under global change. The soil in which plants have grown generally contain AMF spores and a delicate web of hyphae; however, the structures within roots are only visible under the microscope. As expected, the identification of AMF is under changes, being necessary to understand the occurrence of all species.

AMF have major effects on plant growth in a number of ways, the best known being by aiding phosphorus uptake (Smith and Read, 2008). It is known that different AMF communities can be found in degraded vs. pristine forest (de Souza et al. 2013, Haug et al. 2010, Pagano et al. 2011, Soteras et al. 2013). Moreover, different AMF communities can be present in different land use systems (Leal et al. 2009, Sene et al. 2012, Stürmer and Siqueira 2011) and different vegetation types (Pagano et al. 2011, Velazquez et al. 2013).

Which Factors Most Affect the AMF Community

Currently, to identify which factors most affect the AMF community is a challenge. It seems to be a preferred association with particular plant species and families (Torrecillas et al. 2012) or different functional plant groups (Chagnon et al. 2013, Pagano et al. 2013, Zangaro et al. 2013). However, in the last years the characteristics of soils (including the conditions of agriculture) have been more examined and better related to AMF occurrence (Jansa et al. 2014, Wetzel 2014). To accomplish this goal, experimental measurements of AMF community are linked to the properties of soils and sites. However, standardized methodologies for analyses of soil properties are needed in order to assist more detailed and realistic comparisons across studies (Jansa et al. 2014).

Large variation of soil properties can be noted between sampling sites, with pH ranging from acidic to alkaline, and texture from heavily clayey to very sandy. Jansa et al. (2014) stressed that soil and geography are more important determinants of indigenous AMF communities than management practices. However, much research is needed to achieve best conclusive results. For many chemical soil properties, the range of their variation across sites reached two orders of magnitude (Table 1).

Other known limitations are the low comparability of individual studies due to the major confounding effects of dissimilar sample preparation, irregularities associated with PCR protocols and the use of inexact abundance counts (See Jansa et al. 2014). In addition, the morphological identification requires a careful eye to spot AMF spores from different soil samples and to identify them.

It is known that the site geography (altitude and large geographical distance) intensely affect AMF communities (Jansa et al. 2014, Pagano et al. 2012, Pagano 2012). Land management effect on the AMF communities is very useful for us, compared to the other predictors, which could be due to the rarity of highly intensive and unsustainable land management. However, it is difficult to identify any taxon suitable as an indicator of land use along extensive coverage of large geographical and soil gradients (Jansa et al. 2014). Recently, Kowalchuk (2012) draw attention to the possibility that stimulation of AMF under elevated atmospheric CO_2 conditions result in a net loss of carbon from the soil. He stressed the need of interdisciplinary methodologies together with robust field and glasshouse experiments, to manipulate soil nitrogen, organic matter quality, and field management, to understand the balance of soil decomposition processes.

Arbuscular Mycorrhizal Structures

In her review Treseder (2013) concluded that percent root length colonized by AMF structures as measure of AMF abundance is generally used. Additionally, some investigators report percent root length colonized by arbuscules, but these structures are more ephemeral than other intraradical AMF structures, and they can be difficult to observe, so this metric is rarely informed. Mainly, it will depend of the recovered samples which are related to the time of sampling or plant host, and sometimes you can found very clearly these structures and thus, you can annotate their percent.

Total root length colonized by AMF in a specific plant species may be a closer estimate of AMF abundance within a plant host. Total root length colonized by AMF in per-ground area basis, gives an index of ecosystemic abundance.

Table 1. Journal articles dealing with arbuscular mycorrhizas structures in pristine or restored lands

Key words	Number of journal articles
Arbuscular mycorrhizal structures	958
AMF + Restoration	117
AMF + Natural forest	31
Arbuscular mycorrhizal arbuscules	393
Arbuscular mycorrhizal vesicles	294
Arbuscular mycorrhizal extra-radical hyphae	19

Database survey conducted on May 2014 (SCOPUS); AMF = arbuscular mycorrhizal fungi.

The extraradical hyphae (responsible for nutrient uptake by AMF from soil) give the hyphal length in soils (index of AMF biomass), but it is not frequently reported (Table 1). Among the studies, the extra-radical mycelia and spore formation of common species such as *Glomus intraradices* was observed in vitro monoxenic culture system showing the mechanism of survival strategy (Lee 2011). The detailed hyphal networks and spores formation can be registered.

Quantitative PCR of AMF-specific DNA from roots or soil was increasingly reported, in spite of the fact that it is relatively technically demanding and expensive. Gorzelak et al. (2012) highlighted the usefulness of molecular approaches for AMF community ecology. Finally, there are various index of the standing biomass of AMF, but they are not necessarily a measure of their activity or function (reviewed in Smith and Read 2008). More recently, Jansa et al. (2014) stressed the difficulty to know information on land use and its history and the possibility of AMF serve as indicators of land use. For example, a high abundance of *Funneliformis mosseae* could probably indicate previous mineral fertilizer applications. However, to developing a universal AMF indicator for soil quality or land use is still difficult. Likewise, absence of some AMF such as specialist taxa could be more related more to the land management practices. For instance, AMF are good indicators of soil health (Pagano et al. 2011).

With respect to biochar addition as soil amendment, there is a rapidly accumulating body of literature in this regard (e.g., Thies and Rillig 2009). However, there are also potentially important differences that necessitate testing biochar for potential negative effects. Whether or not AMF is related to the benefits of biochar effect is a matter of current discussion. Recently, Pagano and Jorio (2014) compiled information related to biochar / compost and AMF contributing to current information on mycorrhizas with respect to the benefits of biochar/ compost amendments.

Arbuscular Mycorrhizal Fungal TAXA, Morphology, and Activity

AMF produce big spores (Figure 1), as even the largest, the 'giant spores' of *Gigaspora* and some *Scutellospora* are less than half a millimeter across. It is remarkable that AM fungal taxa vary in their contribution to nutrient uptake by plants (Munkvold et al. 2004, Smith and Smith 2011). It is known that

Scutellospora and *Gigaspora,* form more extensive extraradical hyphae, resulting in better nutrient acquisition (Hart and Reader 2002, Treseder 2005, de Souza et al. 2005, Maherali and Klironomos 2007, Powell et al. 2009). As the differences in the extent of root and soil colonization by AMF evolved early in the diversification of AM fungi, different strategies seem to be influenced by functional interactions with host plants (Powell et al. 2009).

On the other hand, *Acaulospora, Glomus, Funneliformis* and *Rhizophagus* produce less extraradical hyphal biomass, and may serve as poorer mutualists (Hart and Reader 2002, Maherali and Klironomos 2007, Powell et al. 2009). Anyway, the extraradical hyphal biomass vary among AMF taxa (Klironomos 2003), but the benefits of different AMF taxa for host plants were not very decisive (Hart and Reader 2002).

However, *Rhizophagus irregularis* or *F. mosseae* seem to be globally distributed and abundant in most ecosystems, whereas others seem to be more restricted (Giovannetti et al. 2003, Opik et al. 2006, 2013, Pagano et al. 2013). Additionally, it is known that distinct mycorrhizal taxa at a single site do usually not exceed 30 (Jansa et al. 2002, Oehl et al. 2003, Dumbrell et al. 2010, Kivlin et al. 2011). However, in pristine sites a higher number can be observed. In Brazil, Pagano et al. (2009) have showed a significant AMF diversity in preserved sites. They obtained the mycorrhizal status for 33 terrestrial plants and ferns investigated both in riparian and dry forests (Pagano 2011, Pagano and Araújo 2011). This information is useful for prioritize sites for conservation such as undisturbed forests, as well as reforested sites such as agroforestry systems and riparian farm forest (previously degraded by cattle). Several studies showed that AMF abundance, biomass and diversity are more important drivers of management-induced changes in aggregate stability and soil nutrients; moreover, increased soil macroporosity in relation to more disturbed sites speaks in favor of an increasing soil quality because macropores enable rapid drainage through the soil profile promoting a better tree growth.

Due to the relatively low diversity of AMF taxa, the different preferences of the individual taxa for environmental properties, and their global distribution, AMF are promising candidates as bioindicators of soil quality, soil health, management history, soil pollution and pesticides application. This fact is crucial for understanding and supporting the sustainable use of soils and ecosystem resources (Fitter et al. 2004, Jansa et al. 2014, Verbruggen et al. 2013).

Some members of the order Glomerales (e.g. *Funneliformis, Rhizophagus and Claroideoglomus*) are very abundant in agroecosystems around the globe

(Helgason et al. 1998, Mathimaran et al. 2005, Gai et al. 2009, Galvan et al. 2009, Rosendahl et al. 2009, Lumini et al. 2010, Moebius-Clune et al. 2013). Moreover the most common species usually are *F. mosseae*, *R. irregularis* and *Claroideoglomus claroideum* (Jansa et al. 2014). Recently, Wetzel et al. (2014) investigated the impact of tillage on AM fungal communities under temperate climate conditions. They showed a considerable better differentiation of fungal taxa and a more sensitive detection of changes in community composition and diversity by using morphological spore identification compared to molecular methods.

Hernandez and Allen (2013) have observed the AMF activity throughout a diurnal pattern which remained unknown. They noted that growth and dieback events often occurred simultaneously and were coupled with soil temperature and moisture, suggesting a rapid acclimation of the external phase of AM fungi to the immediate environment. They concluded that changes in the environmental conditions and variability of the mycorrhizosphere may alter the diurnal patterns of productivity of AM fungi, thereby modifying soil carbon sequestration, nutrient cycling and benefits for host plant.

In 1991, Read reviewed important mycorrhizal knowledge in a special issue of Experientia. In the last years approximately 12,319 peer reviewed journal papers on mycorrhizal structures were reported. Among them, 958 focused on AMF structures (Table 1). Some studies showed that perennial plant species harbor few number of AMF sequence types (eight phylotypes), while other species can be colonized by 13 to 20 fungal types (Torrecillas et al. 2012). Other reports suggest that AMF associations with plants may occur at a higher level, such as the plant functional type or ecological group (Opik et al. 2009), and not necessarily at the species level (van der Heijden and Horton 2009).

There is a need for more scientific information on forest practices in tropical biomes considering the urgent demand for their forest restoration (Aronson et al. 2011). Pagano et al. (2012) compiled information on soil microbes associated to natural forests from South America.

Inoculation with AMF can benefit plant survival in restoration experiments (Matias et al. 2009). However, further studies are needed to understand the role of soil microbiota in those ecosystems. In native forest, Zangaro (2012) has showed that the AMF root colonization of native trees and the spore density in the soil decrease with the advance of ecological succession. However, other studies reported increased occurrence of mycorrhizal tree species in mature forest (Huante et al. 1993, Janos 1980). Allen et al. (2003) concluded that there are few field observations and at the seedling phase. They also showed no specific beneficial relationship between late-seral AMF inocula and late-seral tree species. This divergence in the obtained results needs urgent research to understand the influence of AMF upon tropical forests.

Particular AMF communities were found in different vegetation types in Brazil (de Souza et al. 2013, Freitas et al. 2013, Pagano et al. 2011) and in Argentina (Becerra and Zak 2011, Pagano et al. 2012, Velazquez et al. 2013) mainly because of their association with particular plant species or plant families. The increasing number of new AMF species associated with currently studied plants (Błaszkowski et al. 2013, Carvalho et al. 2012, Velazquez et al. 2008) point out the need to increase research. Some reports showed the same AMF species in different plant types (Freitas et al. 2013), whereas different functional plant groups seem to harbor different AMF identity (Chagnon et al. 2013, Pagano et al. 2013, Zangaro et al. 2013). Studies on tree species used in forestation are incipient and little is known about the characteristics of growth of native species (Pagano and Araújo 2011). The selection of fast-growing legume trees and their mycorrhizal and rhizobial symbioses (de Faria et al. 2010) has resulted in successful indication of more than 87 legume species for reforestation and/or reclamation processes in Brazil (Chaer et al. 2011, Pagano et al. 2010). Inoculated legume species can influence plant growth, seedling survival, rhizospheric effect, etc. Agroforestry systems reporting good alternatives to their agropecuary uses (Paterniani et al. 2010) or using native species plantations (Nascimento et al. 2012) can also be better alternatives for environmental exploitation. With regard to riparian sites, plantations of high-diversity native tree species (Rodrigues et al. 2011) and compiled information on AMF was reported by Pagano (2012, 2013). Most of research on mycorrhizas in Brazilian forests has been made by teams at different Universities. Results obtained during the last years of using native and non-native trees for reforestation in Brazil indicate that mycorrhizas are a common component in forest (Cardoso et al. 2003, Pagano 2012). Pagano and Araújo (2011) compiled the mycorrhizal status of selected tree species studied in Northeast of Brazil, some of them endangered or near threatened trees. Most of the trees were colonized by AMF and were associated with different AMF families, which have different life strategies (Chagnon et al. 2013). Moreover, some empirical observations have shown the natural AMF colonization capacity of soils from a tropical dry forest (Covacevich and Berbara 2011), revealing the sensitivity of AMF to changing environment factors. Spore viability can be poor in dry forests converted into subsistence agriculture or woody production (Lima et al. 2007). Recent reports showed interest in reforestation with native and also exotic tree species, mainly *Eucalyptus*. Mixed plantations such as agroforestry systems with eucalypt and local native species can also provide wood supply for local communities (Pagano et al. 2010) or supply food, forage and N to the soil (for habitat, shade and soil quality enhancement) (Foroughbakhch et al. 2010). Thus, multipurpose trees, most of them leguminous, are extremely useful.

Research on Glomalin or Glomalin Related Soil Protein, a glycoprotein produced primarily by AMF and restricted to hyphae walls (Wright and Upadhyaya, 1996) is slowly increasing. Nowadays 227 scientific papers are reported in the SCOPUS database. Glomalin is a thermo-stable, operationally defined soil protein, having important influence to C pool in soils. The term glomalin was originally thought to be produced only by *Glomus* fungi (Woignier et al. 2014).

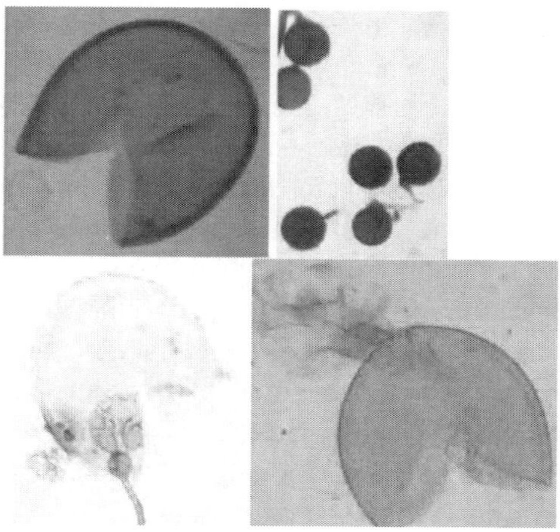

Figure 1. Morphological structures of arbuscular mycorrhizas established outside roots. Clockwise from upper left: spores of *Gigaspora, Septoglomus, Acaulospora* and *Scutellospora*. Photo credit - M. Pagano.

It is believed that AM hyphae decompose and liberate glomalin residue in the soil (Treseder and Turner, 2007). The protein is extracted against soil by autoclaving in citrate solutions. Today many limitations were postulated with regard to the fungal origin of glomalin and its identity. Moreover, other proteins resist autoclaving and can be co-extracted with humic substances that can interfere with glomalin assay and quantification by the colorimetric method of Bradford. Interest in glomalin is observed worldwide, but researchers from Spain who work within semiarid ecosystems focused in this research (Lozano et al. 2014). Usually, their easy evaluation and little soil demand (generally 1g) support their assessment.

CONCLUSION

In the introduction to this chapter, the ecosystem services of AMF were described as related to plant and soil environments. Preserving their biodiversity will help to maintain their multifunctionality (C storage, productivity, nutrients conservation), which can be decisive to buffer the negative effects of global change. Throughout the chapter, the appreciation that most plants associate with different functional types of mycorrhizas and that the most common AMF structures are generally reported, were pointed out. As mycorrhizal fungi respond to disturbances, while other organisms do not, they have greater potential as bioindicators but, further studies are required to understand these microorganisms. Finally, this chapter showed that trees themselves can act as a sink of CO_2 and can increase soil organic matter but their associated AMF require urgent research.

ACKNOWLEDGMENTS

Marcela Pagano is grateful to Minas Gerais State Agency for Research and Development (FAPEMIG) Brazil (Process 311/07), to CAPES for Post-doctoral scholarships granted from 2003 to 2014, including CAPES-PNPD Process N° 23038.007147/2011-60 and MCTI/CNPQ Nº 14/2012 (Universal Grant) from Conselho Nacional de Desenvolvimento Científico e Tecnológico (CNPq).

REFERENCES

Allen, MF; Swenson, W; Querejeta, JI; Egerton-Warbunton, LM; Treseder, KK. Ecology of mycorrhizae: A conceptual framework for complex interactions among plants and fungi. *Annual Review of Phytopathology*, 2003, 41, 271-303.

Aronson, J; Brancalion, PHS; Durigan, G; Rodrigues, RR; Engel, VL; et al. What Role Should Government Regulation Play in Ecological.

Restoration? Ongoing Debate in São Paulo State, Brazil. *Restoration Ecology*, 2011, 19, 6, 690-695.

Becerra, AG; Zak, MR. The ectomycorrhizal symbiosis in South America: Morphology, colonization, and Diversity. In: Rai M, Varma A, editors. *Diversity and Biotechnology of ectomycorrhiza*, Soil Biology, Springer-Verlag Berlin Heidelberg, 2011.

Błaszkowski, J; Chwat, G; Kovács, GM; Gáspár, BK; Ryszka, P; et al. *Septoglomus fuscum* and *S. furcatum*, two new species of arbuscular mycorrhizal fungi (Glomeromycota). *Mycologia*, 2013, 104, 12-127.

Borie, F; Rubio, R; Rouanet, JL; Morales, A; Borie, G; Rojas, C. Effects of tillage systems on soil characteristics, glomalin and mycorrhizal propagules in a Chilean Ultisol. *Soil & Tillage Research*, 2006, 88, 253-261.

Cardoso, IM, Kuyper, TW. Mycorrhizas and tropical soil fertility. *Agriculture, Ecosystems & Environment*, 2006, 116, 72-84.

Cardoso, IM; Boddington, CL; Janssen, BH; Oenema, O; Kuyper, TW. Distribution of mycorrhizal fungal spores in soils under agroforestry and monocultural coffee systems in Brazil. *Agroforestry Systems*, 2003, 58, 33-43.

Carvalho, F; de Souza, FA; Carrenho, R; Moreira, FMS; Jesus, EC; Fernandes, GW. The mosaic of habitats in the high-altitude Brazilian rupestrian fields is a hotspot for arbuscular mycorrhizal fungi. *Applied Soil Ecology*, 2012, 52, 9-19.

Chaer, G M; Resende, AS; Campello, EFC; de Faria, SM; Boddey, RM. Nitrogen-fixing legume tree species for the reclamation of severely degraded lands in Brazil. *Tree Physiology*, 2011, 31, 139-149.

Chagnon, P-L; Bradley, RL; Maherali, H; Klironomos, JN. A trait-based framework to understand life history of mycorrhizal fungi. *Trends in Plant Science*, 2013, 18, 9, 484-91.

Covacevich, F; Berbara, RLL. Indigenous arbuscular mycorrhizae in areas with different successional stages at a tropical dry forest biome in Brazil. *African Journal of Microbiology Research*, 2011, 6, 27, 5523-5535.

Covacevich, F; Berbara, RLL. The impact of climate changes on belowground: How the CO_2 increment affects arbuscular mycorrhiza?. In: *Perspectives in Mycorrhizal Research*. Thangadurai E, Hijri M, Busso CA, editors. Bioscience Publications. 2010, 203-211.

Covacevich, F; Echeverría, HE; Pagano, MC. Arbuscular mycorrhizal fungi: essential belowground organisms for earth life but sensitive to a changing environment. *African Journal of Microbiology Research*, 2012, 6, 27, 5523-5535.

de Faria, SM; Campello, EF; Xavier, DF; Boddey, RM. Multi-purpose fast-growing legume trees for smallholders in the tropics and sub-tropics: firewood, fencing and fodder. In: *Comunicado Técnico*. Embrapa Agrobiologia, Seropédica. 2010, pp. 6, http://www.cnpab.embrapa.br/publicacoes/download/cot127.pdf.

De Souza, FA; Dalpé, Y; Declerck, S; de la Providencia, IE; Séjalon-Delmas, N. Life History Strategies in Gigasporaceae: Insight from Monoxenic Culture. In: Declerck S, Strullu DG, Fortin, JA, editors. *In Vitro Culture of Mycorrhizas*. Springer; 2005; 4, 73–91.

de Souza, RG; Silva, DKA; Mello, CMA; Goto, BT; Silva, FSB; Sampaio, EVSB; Maia, LC. Arbuscular mycorrhizal fungi in revegetated mined dunes. *Land Degradation and development*, 2013, 24, 2, 147-155.

Dominati, E; Patterson, M; Mackay, A. A framework for classifying and quantifying the natural capital and ecosystem services of soils. *Ecological Economics*, 2010, 69, 1858-68.

Duarte, EMG; Cardoso, IM; Stijnen, T; Mendonça, MAFC; Coelho, MS; et al. Decomposition and nutrient release in leaves of Atlantic Rainforest tree species used in agroforestry systems. *Agroforestry Systems*, 2013, 87, 835-847.

Dumbrell, AJ; Nelson, M; Helgason, T; Dytham, C; Fitter, AH. Relative roles of niche and neutral processes in structuring a soil microbial community. *The ISME Journal*, 4, 337–345, 2010a.

Fitter, AH; Heinemeyer, A; Husband, R; Olsen, E; Ridgway, KP; Staddon, PL. Global environmental change and the biology of arbuscular mycorrhizas: gaps and challenges. *Can. J. Bot.*, 2004, 82, 1133-1139.

Foroughbakhch, RP; Piñero, JH; Vásquez, MAA; Avila, MLC. Use of multipurpose trees and shrubs in forestry and agroforestry systems in northeastern Mexico. In: LR Kellymore, editor. *Handbook on Agroforestry: Management Practices and Environmental Impact*. Nova Science Publishers, New York. 2010, 325-345.

Freitas, RO; Buscardo, E; Nagy, L; Maciel, ABS; Carrenho, R; Luizão, RCC. Arbuscular mycorrhizal fungal communities along a pedohydrological gradient in a Central Amazonian terra firme forest. *Mycorrhiza*, 2013, 24,1, 21-32.

Gai, JP; Christie, P; Cai, XB; Fan, JQ; Zhang, JL; Feng, G; Li, XL. Occurrence and distribution of arbuscular mycorrhizal fungal species in three types of grassland community of the Tibetan Plateau. *Ecological Research*, 2009, 24, 6, 1345-1350

Galván, GA; Parádi, I; Burger, K; Baar, J; Kuyper, TW; Scholten, OE; Kik, C. Molecular diversity of arbuscular mycorrhizal fungi in onion roots from organic and conventional farming systems in the Netherlands. *Mycorrhiza*, 2009, 19, 5, 317-328.

Giovannetti, M; Sbrana, C; Strani, P; et al. Genetic diversity of isolates of Glomus mosseae from different geographic areas detected by vegetative compatibility testing and biochemical and molecular analysis. *Appl. Environ. Microbiol.*, 2003, 63, 616-624.

Gorzelak, MA; Holland, TC; Xing, X; Hart, MM. Molecular approaches for AM fungal community ecology: A primer. *Journal of Microbiological Methods*, 2012, 90, 108-114.

Hart, MM; Reader, RJ. Taxonomic basis for variation in the colonization strategy of arbuscular mycorrhizal fungi. *New Phytol.*, 153, 335-344, 2002.

Haug, I; Wubet, T; Weiß, M; Aguirre, N; Weber, M; et al. Species-rich but distinct arbuscular mycorrhizal communities in reforestation plots on degraded pastures and in neighboring pristine tropical mountain rain forest. *Tropical Ecology*, 2010, 51, 2, 125-148.

Helgason, T; Daniell, TJ; Husband, R; Fitter, AH; Young, JPW. Ploughing up the wood-wide web? *Nature*, 1998, 394, 431.

Hernandez, RR; Allen, MF. Diurnal patterns of productivity of arbuscular mycorrhizal fungi revealed with the Soil Ecosystem Observatory. *New Phytologist*, 2013, 200, 547-557.

Huante, P; Rincón, E; Allen, EB. Effect of vesicular-arbuscular mycorrhizae on seedling growth of four tree species from the tropical deciduous forest of Mexico. *Mycorrhiza*, 1993, 2, 141-145.

Janos, DP. Mycorrhizae influence tropical succession. *Biotropica*, 1980, 12, 56-64.

Jansa, J; Erb, A; Oberholzer, H; Smilauer, P; Egl, S. Soil and geography are more important determinants of indigenous arbuscular mycorrhizal communities than management practices in Swiss agricultural soils. *Molecular Ecology*, 2014, doi: 10.1111/mec.12706

Jansa, J; Mozafar, A; Anken, T; Ruh, R; Sanders, IR; Frossard, E. Diversity and structure of AMF communities as affected by tillage in a temperate soil *Mycorrhiza*, 2002, 12, 225-234

Kahiluoto, H; Ketoja, E; Vestberg, M. Contribution of arbuscular mycorrhiza to soil quality in contrasting cropping systems. *Agriculture, Ecosystems and Environment*, 2009, 134, 36-45.

Kivlin, SN; Hawkes, CV; Treseder, KK. Global diversity and distribution of arbuscular mycorrhizal fungi. *Soil Biology and Biochemistry*, 2011, 43, 11, 2294-2303.

Kowalchuk, GA. Bad News for Soil Carbon Sequestration? *Science*, 2012, 337, 1049-1050.

Lal, R. Soils and food sufficiency. A review. *Agron. Sustain. Dev.*, 2009, 29,113-133.

Leal, PL; Stürmer, SL; Siqueira, JO. Occurrence and diversity of arbuscular mycorrhizal fungi in trap cultures from soils under different land use systems in the Amazon, Brazil. *Brazilian Journal of Microbiology*, 2009, 40,111-121.

Lee, J. The Distribution of Cytoplasm and Nuclei within the Extra-radical Mycelia in, a Species of Arbuscular Mycorrhizal Fungi. *Mycobiology*, 2011, 39(2), 79-84.

Lima, RLA; Salcedo, IH; Fraga, VS. Propágulos de fungos micorrízicos arbusculares em solos deficientes em fósforo sob diferentes usos da região semi-árida no nordeste do Brasil. *Rev. Brasil. Ciênc. Solo*, 2007, 31, 257-268.

Lozano, E; García-Orenes, F; Bárcenas-Moreno, G; Jiménez-Pinilla, P; Mataix-Solera, J; Arcenegui, V; Morugán -Coronado, A; Mataix-Beneyto, J. Relationships between soil water repellency and microbial community composition under different plant species in a Mediterranean semiarid forest. *J. Hydrol. Hydromech.*, 2014, 62, 2, 101-107.

Lumini, E; Orgiazzi, A; Borriello, R; Bonfante, P; Bianciotto, V Disclosing arbuscular mycorrhizal fungal biodiversity in soil through a land-use gradient using a pyrosequencing approach. *Environmental Microbiology*, 12, 2165-2179, 2010.

Maherali, H; Klironomos, JN. Influence of phylogeny on fungal community assembly and ecosystem functioning. *Science*, 2007, 316, 1746-1748.

Mathimaran, N; Ruh, R; Vullioud, P; Frossard, E; Jansa, J. *Glomus intraradices* dominates arbuscular mycorrhizal communities in a heavy textured agricultural soil. *Mycorrhiza*, 2005, 16:1, 61-66.

Matias, SR; Pagano, MC; Muzzi, FC; Oliveira, CA; Carneiro, AA; et al. Effect of rhizobia, mycorrhizal fungi and phosphate-solubilizing microorganisms in the rhizosphere of native plants used to recover an iron ore area in Brazil. *European Journal of Soil Biology*, 2009, 45, 3, 259-266.

Moebius-Clune, D; Moebius-Clune, BN; van Es, HM; Pawlowska, TE. Arbuscular mycorrhizal fungi associated with a single agronomic plant host across the landscape: Community differentiation along a soil textural gradient. *Soil Biology and Biochemistry*, 2013, 64, 191-199.

Munkvold, L; Kjøller, R; Vestberg, M; Rosendahl, S; Jakobsen, I. High functional diversity within species of arbuscular mycorrhizal fungi. *New Phytologist*, 2004, 164, 357-364.

Nascimento, DF; Leles, PSS; Oliveira Neto, SN; Moreira, RTS; Alonso, JM. Initial growth of six forest tree species in different spacing conditions. *Cerne*, 2012, 18, 1, 159-165.

Nielsen, UN; Ayresa, E; Wall, DH; Bardgett, RD. Soil biodiversity and carbon cycling: A review and synthesis of studies examining diversity-function relationships. *European Journal of Soil Science*, 2011, 62, 105-116.

Nyfeler, D; Huguenin-Elie, O; Suter, M; Frossard, E; Lüscher, A. Grass-legume mixtures can yield more nitrogen than legume pure stands due to mutual stimulation of nitrogen uptake from symbiotic and non-symbiotic sources. *Agriculture, Ecosystems & Environment*, 2011, 140, 1-2, 155-163.

Oehl, F; Sieverding, E; Ineichen, K; Mäder, P; Boller, T; Wiemken, A. Impact of land use intensity on the species diversity of arbuscular mycorrhizal fungi in agroecosystems of Central Europe. *Appl Environ Microbiol.*, 2003, 69 (5), 2816-24.

Opik, M; Moora, M; Liira, J; Zobel, M. Composition of rootcolonizing arbuscular mycorrhizal fungal communities in different ecosystems around the globe. *Journal of Ecology*, 94, 778-790, 2006.

Öpik, M; Zobel, M; Cantero, JJ; Davison, J; Facelli, JM; Hiiesalu, I; Jairus, T; Kalwij, JM; Koorem, K; Leal, ME; Liira, J; Metsis, M; Neshataeva, V; Paal, J; Phosri, C; Põlme, S; Reier, Ü; Saks, Ü; Schimann, H; Thiéry, O; Vasar, M; Moora, M. Global sampling of plant roots expands the described molecular diversity of arbuscular mycorrhizal fungi. *Mycorrhiza*, 2013, 23, 5,411-30.

Pagano, MC, editor. *Mycorrhiza: Occurrence in Natural and Restored Environments*, Nova Science Publishers, New York, USA, 2012.

Pagano, MC. Plant and soil biota: Crucial for mitigating climate change. In: Interactions of Forests, Climate, Water Resources, and Humans in a Changing Environment. *British Journal of Environment and Climate Change*, SCIENCEDOMAIN international, London, UK, 2013, 3 (2).

Pagano, MC; Araújo, FS. Semiarid Vegetation in Brazil: Biodiversity, Impacts And Management. In: Degenovine KM, editor. *Semi-Arid Environments: Agriculture, Water Supply and Vegetation.* Nova Science Publishers, New York, 2011, 99-114.

Pagano, MC; Cabello, MN. Importance of mycorrhizal interactions for reforestation. *ISRN Ecology*, Hindawi Publishing Corporation. ISSN: 2090-4614, e-ISSN: 2090-4622, doi:10.5402. 2011.

Pagano, MC; Cabello, MN; Scotti, MR. Agroforestry in Dry Forest, Brazil: Mycorrhizal Fungi Potential. In: Kellymore LR, editor. *Handbook on Agroforestry: Management Practices and Environmental Impact.* Nova Science Publishers, New York. 2010, 367-388.

Pagano, MC; Dhar, PP. Mycorrhizas associated with forest under climate change. In: Biology of Bioactive Compounds. Lac S, Lucas-Borja M, editors. Nova Science, New York, USA, 2014, 119-134.

Pagano, MC; Jorio, A. The contributions of mycorrhizal fungi. In: Microbial Bioresources. Gupta V. K., Thangdurai D., Sharma G.D, editors, CABI, UK, 2014 (in press).

Pagano, MC; Lugo, M; Araújo, F; Ferrero, M; Menoyo, E; Steinaker, D. Native species for restoration and conservation of biodiversity in South America. In: Marín L and Kovač D, editors, *Native Species: Identification, Conservation and Restoration*, Nova Science Publishers, New York, USA, 2012, 1-55.

Pagano, MC; Schalamuk, S; Cabello, MN. Arbuscular mycorrhizal parameters and indicators of soil health and functioning: Applications for agricultural and agroforestal systems. In: Miransari M, editor. *Soil Microbes and Environmental Health*, Nova Science Publishers, New York, USA., 2011, 267-276.

Pagano, MC; Scotti, MR. A preliminary survey of the occurrence of arbuscular mycorrhizas in *Paepalanthus bromelioides* and *Bulbostylis* sp. in Brazil. *Journal of Applied Mycology*, 21(1), 1-10.

Pagano, MC; Utida, MK; Gomes, EA; Marriel, IE; Cabello, MN; Scotti, MR. Plant-type dependent changes in arbuscular mycorrhizal communities as soil quality indicator in semi-arid Brazil. *Ecological Indicators*, 2011, 11, 2, 643-650.

Pagano, MC; Zandavalli, RB; Araújo, FS. Biodiversity of arbuscular mycorrhizas in three vegetation types from the semiarid of Ceará State, Brazil. *Applied Soil Ecology*, 2013, 67, 37-46.

Paterniani, MLS; Pardini, A; Pratesi, V. Integration of native genetic resources in Brazilian agroforestry systems. In: Kellymore LR, editor. *Handbook on*

Agroforestry: Management Practices and Environmental Impact. Nova Science Publishers, New York, 2010, 301-324.

Powell, JR; Parrent, JL; Hart, MM; Klironomos, JN; Rillig, MC; Maherali, H. Phylogenetic trait conservatism and the evolution of functional trade-offs in arbuscular mycorrhizal fungi. *Proceedings of the Royal Society B-Biological Sciences*, 2009, 276, 4237-4245.

Raupach, RM; Marland, G; Ciais, P; Quéré, C; Canadell, JG; et al. Global and regional drivers of accelerating CO_2 emissions. *PNAS*, 2007, 104, 10288-10293.

Read, DJ. Mycorrhizas in ecosystems. *Experientia*, 1991, 47, 4.

Redondo-Brenes, A; Montagnini, F. Contribution of homegardens, agrosilvopastoral systems, and other human-dominated land-use types to the avian diversity of a biological corridor in Costa Rica. In: Kellymore LR, editor. *Handbook on Agroforestry: Management Practices and Environmental Impact.* Nova Science Publishers, New York, 2010, 185-224.

Rodrigues, RR; Gandolfi, S; Nave, AG; Aronson, J; Barreto, TE; Vidal, CY; Brancalion, PHS. Large-scale ecological restoration of high-diversity tropical forests in SE Brazil *Forest Ecology and Management*, 2011, 261, 10, 1605-1613.

Rosendahl, S; Mcgee, P; Morton, JB. Lack of global population genetic differentiation in the arbuscular mycorrhizal fungus *Glomus mosseae* suggests a recent range expansion which may have coincided with the spread of agricultura. *Molecular Ecology*, 2009, 18, 20, 4316-4329.

Sala, OE; Chapin III, FS; Armesto, JJ; Berlow, E; Bloomfield, J; Dirzo, R. Global Biodiversity Scenarios for the Year 2100. *Science*, 2000, 287, 1770-1774.

Sene, G; Samba-Mbaye , R; Thiao, M; Khasa, D; Kane, A; et al. The abundance and diversity of legume-nodulating rhizobia and arbuscular mycorrhizal fungal communities in soil samples from deforested and man-made forest systems in a semiarid Sahel region in Senegal. *European Journal of Soil Biology*, 2012, 52, 30-40.

Simard, S; Austin, ME. The role of mycorrhizas in forest soil stability with climate change. In: *Climate change and variability.* (Ed.) Simard, S., Austin, M. E., 2010, 275-302.

Smith, SE; Read, DJ. Mycorrhizal Symbiosis. New York: Elsevier, 2008.

Smith, SE; Smith, FA. Roles of arbuscular mycorrhizas in plant nutrition and growth: new paradigms from cellular to ecosystems scales. *Annu. Rev. Plant Biol.*, 2011, 63, 227-250

Soteras, F; Renison, D; Becerra, AG. Growth response, phosphorus content and root colonization of *Polylepis australis* Bitt. Seedlings inoculated with different soil types. *New forests*, 44, 4, 577-589, 2013.

Stürmer, SL; Siqueira, JO. Species richness and spore abundance of arbuscular mycorrhizal fungi across distinct land uses in Western Brazilian Amazon. *Mycorrhiza*, 2011, 21, 255-267.

Thies, JE; Rillig, M. Characteristics of biochar: biological properties. In: Lehmann, J., Joseph, S. (Eds.), Biochar for Environmental Management: Science and Technology. Earthscan, London, 2009, 85-105.

Torrecillas, E; Alguacil, MM; Roldán, A. Differences in the AMFdiversity in soil and roots between two annual and perennial gramineous plants co-occurring in a Mediterranean, semiarid degraded area. *Plant Soil*, 2012, 354, 97-106.

Treseder KK; Allen MF; Ruess RW; Pregitzer KS; Hendrick RL. Lifespans of fungal rhizomorphs under nitrogen fertilization in a pinyon–juniper woodland. *Plant and Soil*, 2005, 270, 249-255.

Treseder, KK. The extent of mycorrhizal colonization of roots and its influence on plant growth and phosphorus content. *Plant Soil*, 2013, 371,1-13.

Treseder, KK; Turner, KM. Glomalin in ecosystems. *Soil Sci. Soc. Am. J.*, 2007, 71, 1257-1266.

Tundisi, JG; Tundisi, TM. Impactos potenciais das alterações do Código Florestal nos recursos hídricos. *Biota Neotrop.*, 2010, 10 (4), 67-75.

van der Heijden, MGA; Horton, TR. Socialism in soil? The importance of mycorrhizal fungal networks for facilitation in natural ecosystems. *Journal of Ecology*, 2009, 97, 6, 1139-1150.

van der Wal, A; Geydan, TD; Kuyper, TW; de Boer, W. A thready affair: linking fungal diversity and community dynamics to terrestrial decomposition processes. *FEMS Microbiol. Rev.*, 2012, 1-18.

van Groenigen, K; Xuan Qi, J; Craig W, Osenberg; Yiqi, Luo; Hungate, BA. Faster Decomposition Under Increased Atmospheric CO_2 Limits Soil Carbon Storage. *SCIENCE*, 2014, 344,508-509.

Velazquez, MS; Cabello, M; Irrazabal, G; Godeas, A. Acaulosporaceae from El Palmar National Park, Entre Ríos, Argentina. *Mycotaxon*, 2008, 103, 171-187.

Velazquez, MS; Cabello, MN; Barrera, M. Composition and structure of arbuscular-mycorrhizal comunnities in El Palmar National Park, Argentina. *Mycologia*, 2013, 105, 3, 509-520.

Verbruggen, E; van der Heijden, MGA; Rillig, MC; Toby Kiers, E. Mycorrhizal fungal establishment in agricultural soils: factors determining inoculation success. *New Phytologist*, 2013, 197, 4, 1104-1109.

Wall, D H; Nielsen, U N. Biodiversity and Ecosystem Services: Is It the Same Below Ground? *Nature Education Knowledge*, 2012, 3(12), 8.

Wang, B; Qiu, YL. Phylogenetic distribution and evolution of mycorrhizas in land plants. *Mycorrhiza*, 2006, 16, 299-363.

Warnock, DD; Lehmann, J; Kuyper, TW; Rillig, MC. Mycorrhizal responses to biochar in soil – concepts and mechanisms. *Plant Soil*, 2007, 300, 9-20.

Wetzel, K; Silva, G; Matczinski, U; Oehl, F; Fester, T. Superior differentiation of arbuscular mycorrhizal fungal communities from till and no-till plots by morphological spore identification when compared to T-RFLP. *Soil Biology & Biochemistry*, 2014, 72, 88-96.

Woignier, T; Etcheverri, P; Borie, F; Quiquampoix, H; Staunton, S. Role of allophanes in the accumulation of glomalin-related soil protein in tropical soils (Martinique, French West Indies). *European Journal of Soil Science*, 2014, 1-8.

Wright, SF; Upadhyaya, A. Extraction of an abundant and unusual protein from soil and comparison with hyphal protein of arbuscular mycorrhizal fungi. *Soil Science*, 1996, 161, 575–586.

Zangaro, W. Arbuscular mycorrhizas and their importance for tropical forest formation in Brazil. In: *Mycorrhiza: Occurrence in Natural and Restored Environments*, USA. Pagano MC, editor, 2012, 49-73.

Zangaro, W; Rostirola, LV; de Souza, PB; Alves, AR; Lescano, LEAM; et al. Root colonization and spore abundance of arbuscular mycorrhizal fungi in distinct successional stages from an Atlantic rainforest biome in southern Brazil. *Mycorrhiza*, 2013, 23, 221-233.

In: Mycorrhizas
Editor: Elaine Warwick

ISBN: 978-1-63321-436-1

Chapter 3

FUNCTIONS OF MYCORRHIZAE

Tzi Bun Ng *
and Charlene Cheuk Wing Ng
School of Biomedical Sciences, Faculty of Medicine,
The Chinese University of Hong Kong,
New Territories, Hong Kong,China

ABSTRACT

Mycorrhizae represent a symbiotic relationship between a fungus and plant roots. The fungus colonizes the roots of the host plant, either intracellularly in arbuscular mycorrhizal fungi, or extracellularly in ectomycorrhizal fungi. Host plants enhance the defense of mycorrhizal fungi against fungivores. Mycorrhizae mediate host plant-insect interactions.

Arbuscular mycorrhizal exudates are important to the formation of lateral roots, and in the absorption and accumulation of elements including manganese, phosphorus, potassium, and zinc. Mycorrhizae alter nitrogen acquisition and improve plant growth. Organic nitrogen is more accessible to etomycorrhizal and ericoid mycorrhizal fungi than arbuscular mycorrhizal fungi because of the presence of enzymes capable of degrading nitrogen- containing molecules in the former type of fungi. Ectomycorrhizal fungi are involved in decomposing and mobilizing nitrogen from organic matter-protein complexes in forest soils.

* Corresponding author, Email: b021770@mailserv.cuhk.edu.hk, fax: 852 26035123, phone: 852 39436872.

Ectomycorrhizae induce changes in the surface structure and composition of soil colloids.

Mycorrhizae are involved in phytoremediation of soil contaminated with heavy metals. They also enhance plant tolerance to adverse conditions, decrease aluminium phytotoxicity in acidic soils, and improve plant antioxidant systems and tolerance to salinity.

Keywords: Mycorrhizae, functions, host plant, phytoremediation

INTRODUCTION

Mycorrhizae represent a symbiotic relationship between a fungus and plant roots. The fungus colonizes the roots of the host plant, either intracellularly in arbuscular mycorrhizal fungi, or extracellularly in ectomycorrhizal fungi.

Arbuscular mycorrhiza play a crucial role in augmenting plant resistance to stressful conditions like diseases, drought, heavy metal toxicity, high temperatures, salinity, and waterlogging. They are also important in enhancing decomposition of toxic organic substances, in remediating polluted environments, and in keeping a stable ecosystem and sustainable productivity. The structural features of arbuscular mycorrhiza regulate mycorrhizal formation in the root system, and are intimately associated (Tian et al., 2013).

In the following, the various functions of mycorrhizae are discussed.

HOST PLANTS INCREASE DEFENSE OF MYCORRHIZAL FUNGI AGAINST FUNGIVORES

Catalpol, a metabolite of *Plantago lanceolata,* was detected in arbuscular mycorrhizal fungal hyphae in host plants which had been exposed to fungivorous springtails. Catalpol was found in the fungal hyphae only when there were springtails.

The data indicate a transfer of metabolites from host plants to fungi to reinforce fungal defense against their predators, the fungivorous springtails (Duhamel et al., 2013).

MYCORRHIZAE MEDIATE HOST PLANT-INSECT INTERACTIONS

Root-associated microorganisms may elicit physiological changes in the host plant that affect interactions between plants and aboveground insects at various trophic levels. Above ground, plants are subject to incessant assaults of herbivorous insects and develop responses that also exert effects on microorganisms underground. Very often, both ecological and mechanistic investigations have emphasized on exploring these under- and above-ground interactions using simplified systems involving both single microorganism and herbivorous species, which is different from the naturally occurring interactions. In their review, Pangesti et al. (2013) dealt with at length how colonization by either a single non-pathogenic microbial species or a group of underground microorganisms influenced growth and defense of the plants, how this affected plant interactions with insects at various trophic levels above ground, and how plant reactions to herbivorous insects influenced interactions of plants with soil-borne non-pathogenic microorganisms. The part played by plant hormones in coordinating plant growth and defense against herbivorous insects and in interactions with non-pathogenic soil microorganisms was also discussed (Pangesti et al., 2013).

Barber et al. (2013) carried out a greenhouse study in which they innoculated cucumber with arbuscular mycorrhizal fungi from conventional farms, organic farms, and a commercial arbuscular mycorrhizal fungus inoculum. Growth and plant characteristics considered to be important predictors of resistance to herbivorous insects and attraction of pollinators were examined. Some leaf and root characteristics and flower production were altered by the arbuscular mycorrhizal fungus inoculum. Both treatment with conventional fungi and treatment with organic arbuscular mycorrhizal fungi reduced foliage phosphorus content but raised sodium content in comparison with control and commercial arbuscular mycorrhizal fungi. The fungal treatments were devoid of an effect on foliage defense. However, in comparison with organic arbuscular mycorrhizal fungi, conventional fungi elevated the concentration of the defense chemical cucurbitacin C in the root. Arbuscular mycorrhizal fungi from organic as well as conventional farms had reduced production of flowers compared with that from commercial and control treatments, and were thus less attractive to pollinators resulting in a decline in plant reproduction. Arbuscular mycorrhizal fungi from both types of farm inhibited seed germination, without a major impact on plant growth. The data indicate that investigations that take into account of only the effects of arbuscular mycorrhizal fungi on growth may neglect changes in plant characteristics which can affect interactions, and thus the yield, on farms. Hence detailed evaluations of mycorrhizal influence in a complex community, incorporating multi-species interactions both above the soil surface and underground, are warranted (Barber et al., 2013).

Regulation of Root Morphogenesis in Host Plants

Arbuscular mycorrhizae are the basis of a symbiotic relationship between the host plant root and fungus that leads to better acquisition of phosphate by the plant and modifies the physiology and development of host plants. More extensive branching is a general characteristic of arbuscular mycorrhiza roots, and is a way to add suitable colonization sites. Fungal exudates are important to the formation of lateral roots. The enhanced phosphate content of the plants, and alterations in carbohydrate delivery to the roots, plant hormone concentration, transport and sensitivity, all contribute to more extensive root branching. Arbuscular mycorrhiza colonization and a high phosphate level upregulate concentrations of auxin and cytokinin and downregulate concentrations of ethylene and strigolactone. Cytokinins regulate the root: shoot biomass ratio whereas the other hormones regulate root morphogenesis. Ethylene and strigolactones suppress formation of lateral roots in the non-colonized, phosphate-starved plants. On the other hand, auxins induce formation of lateral roots in colonized plants, or in plants grown under high phosphate conditions (Fusconi, 2014).

Effects on Elemental Distribution in Roots

Orłowska et al. (2013) investigated the impact of arbuscular mycorrhizal fungi on the distribution and concentration of elements in roots of the nickel-hyperaccumulating plant *Berkheya coddii*. Manganese, phosphorus, potassium, and zinc accumulated in the cortical layer of lateral roots of inoculated plants, signifying the vital part played by arbuscular mycorrhizal fungi in the absorption and accumulation of these elements. Elevated levels of copper, iron, nickel, phosphorus, potassium, and zinc in the vascular stele in roots of inoculated plants compared with the uninoculated plants imply elemental translocation with increased efficiency to parts of the plant above ground (Orłowska et al., 2013).

Ectomycorrhiza-Induced Changes in Surface Structure and Composition of Soil Colloids

Treatment of dark brown forest soil (a good loam), from either the surface or deep soils, with ectomycorrhizae (*Gomphidius viscidus*) resulted in physical absorption of covering materials on the colloid surface. Consequently the crystallinity of variable soil minerals (hydromica, kaolinite, and quartz) increased by 40-300% and the amount of variable functional groups (C=O stretching, C-H stretching, O-H stretching and bending) rose by 3-26%. However, fungus treatment of saline-alkali soil (heavily degraded soil) produced a 12-35% decline in the majority of functional groups, 15-55% reduction in crystallinity of the majority of soil minerals but enlargement of their grain size, and increase in soil particle size. These distinct responses reduced the carbon to nitrogen, carbon to oxygen and carbon to sulfur ratios in soil colloids from saline-alkali soil, rendering them more similar to those of the good loam in dark brown forest soil (Li et al., 2013).

Effects on Soil Carbon

There is more soil carbon than the summation of atmospheric carbon and carbon present in vegetation. The breakdown of organic matter in soil is often restricted by the availability of nitrogen to microorganisms. Plants, through symbiotic fungi, compete for nitrogen with free-living decomposers. Organic nitrogen is more accessible to etomycorrhizal and ericoid mycorrhizal fungi than arbuscular mycorrhizal fungi because of the presence of enzymes capable of degrading nitrogen-containing molecules in the former fungi. Averill et al.(2014) demonstrated that soil in ecosystems where ectomycorrhizal and ericoid mycorrhizal fungi-associated plants are predominant possesses 70% more carbon per unit nitrogen than soil in ecosystems where arbuscular mycorrhizal fungi -associated plants prevail. The impact of the type of mycorrhizal fungi on soil carbon is much greater than the effects of net primary production, rainfall, temperature, and soil clay content (Averill et al.,2014)..

Decomposing and Mobilizing Nitrogen from Organic Matter-Protein Complexes by Ectomycorrhizal Fungi

Ectomycorrhizal fungi are involved in decomposing and mobilizing nitrogen from organic matter-protein complexes in forest soils. Addition of glucose stimulated nitrogen assimilation and litter material decomposition, and upregulated expression of genes encoding enzymes involved in oxidative degradation of polysaccharides and polyphenols, peptidases, nitrogen transporters and enzymes in carbon and nitrogen metabolic pathways. Addition of ammonium to organic matter had minor effects on the expression of transcripts and litter material decomposition, which took place only in the presence of glucose (Rineau et al., 2013).

Mycorrhizae Alter Nitrogen Acquisition

Orchid roots are coarser, and thus have a much lower surface area to volume ratio than those of other plants. Consequently, orchids are not well equipped for direct uptake of soil nutrients.

Mycorrhizal fungi associated with orchids change the primary nitrogen source taken up by the orchid *Cymbidium goeringii* from nitrate ions to ammonium ions (Wu et al., 2013).

Mycorrhizae Facilitate Nitrogen Utilization of Host

Ectomycorrhial fungi diversity facilitates the host utilization of nitrogen by sequestering nitrogen arising from solutes released early and late breakdown products from a recalcitrant source (Pena et al., 2013).

Mycorrhizae Improve Plant Growth

A greenhouse pot experiment was performed to examine the effects of three kinds of single inoculum (including indigenous mycorrhizas isolated from arsenic mine, *Glomus intraradices* and *G.mosseae*) and two types of mixed inoculum (mixed with indigenous mycorrhizas and either *G. intraradices* or *G. mosseae*) on the growth response of *Cynodon dactylon* (non-hyperaccumulator) and *Pteris vittata* (hyperaccumulator) at three arsenic concentrations (0, 100 and 200 mg/kg). Both mycorrhizal plants displayed a larger biomass, and greater tissue accumulation of phosphorus and nitrogen than the control. Among the mycorrhizal inoculum, the mixed inoculum indigenous mycorrhizae/ *G. mosseae* triggered more mycorrhizal colonization and higher arsenate reductase activity in *P. vittata* than *C. dactylon*, at all three arsenic concentrations. The portion of Paris arbuscular mycorrhizal structure (in colonized roots) together with the highest arsenic translocation factor of 10.2 in *P. vittata* inoculated with indigenous mycorrhizae / *G lomus mosseae* was observed.

Hence indigenous mycorrhizae / *G lomus mosseae* inoculum may be optimal for field inoculation at any contaminated lands as the inoculum was better adapted to changing ambient conditions and thus was beneficial to the host plants (Leung et al., 2013).

Relatedness Among Arbuscular Mycorrhizal Fungi Drives Plant Growth and Intraspecific Fungal Coexistence

Relatedness between arbuscular mycorrhizal fungi promoted plant growth (27% rise in biomass in the case of closely related fungi versus remotely related fungi). Arbuscular mycorrhizal fungi relatedness also promoted intraspecific coexistence of these ecologically important symbionts (Roger et al., 2013).

Influence on Growth, Heat Stress Response and Antioxidative Activity

Maya and Matsubara (2013) examined the effect of the arbuscular mycorrhizal fungus, *Glomus fasciculatum*, on the heat stress responses, antioxidative activity and growth in *Cyclamen persicum*. Cyclamen plants (with or without fungal inoculation) were grown for 3 months in a commercial potting media at 17-20 °C in a greenhouse and then exposed to two thermal conditions in a growth chamber. Initially, plants were grown at 20 °C for one month under conditions without heat stress, followed by 30 °C for another month under conditions of heat-stress. The mycorrhizal symbiosis strikingly improved biomass production and heat stress responses in plants compared to that in the controls. Leaf browning which was severe in control plants was minimal in mycorrhizal plants. The contents of ascorbic acid and polyphenol, activities of antioxidative enzymes including ascorbate peroxidase and superoxide dismutase, and 2,2-diphenyl-1-picrylhydrazyl radical scavenging activity were all heightened in mycorrhizal plants (Maya and Matsubara 2013).

Improving Plant Antioxidant Systems and Tolerance to Salinity

Salinity stress produces secondary oxidative stress in plants and antioxidant capacity correlates with salt tolerance. Arbuscular mycorrhizal fungi enhanced the antioxidant capacity of plants. Maize plants inoculated with three native arbuscular mycorrhizal fungi exhibited enhanced efficiency of photosystem II and stomatal conductance, which decreased photorespiration and reactive oxygen species production.

Hydrogen peroxide accumulation, oxidative damage to lipids and membrane electrolyte leakage in these arbuscular mycorrhizal plants were attenuated compared with nonmycorrhizal plants or plants inoculated with the collection arbuscular mycorrhizal fungi. The activation of antioxidant enzymes such as catalase and superoxide dismutase also explained these effects (Estrada et al., 2013).

Improving the Production and Accumulation of Important Active Ingredients of Medicinal Plants

Arbuscular mycorrhizae improve the production and accumulation of important active ingredients such as alkaloids, terpenes, phenols, and optimizing the composition of different active ingredients in medicinal plants and improving the quality of herbal materials. The symbiosis will improve the total yield and quality of herbal materials (Zeng et al., 2013)

PHYTOREMEDIATION OF SOIL CONTAMINATED WITH HEAVY METALS

Phytoremediation is the deployment of metal-accumulating plants to decrease the detrimental effects of soil contaminants on human health and the environment. Microbially facilitated bioremediation approaches, like phytoextraction or phytostabilization, may enhance the benefits and are potentially valuable for remediation of low and heterogeneously contaminated soil. The plant-microorganism interactions in phytoremediation approaches comprise symbiotic relationships such as mycorrhizae, plant growth-promoting bacteria, or endophytic bacteria (Phieler et al., 2013).

A species of willow characterized by strong natural mycorrhizal colonization (*Salix dasyclados*), and another species of willow possessing weaker natural mycorrhizal colonization (*S. viminalis*), were compared for efficiency of phytoextraction of heavy metals (cadmium, copper, lead and zinc) from contaminated soil (Hrynkiewicz and Baum, 2013). Inoculation with *Amanita muscaria,* an ectomycorrhizal fungus, suppressed the foliage biomass of both inoculated willow clones, and either increased or was without effect on the trunk biomass of *S. dasyclados* and root biomass of *S. viminalis*. The heavy metal levels in the biomass of *S. dasyclados* were elevated compared with those in *S. viminalis* regardless of inoculation with the ectomycorrhizal fungus. Inoculation with *A. muscaria* reduced the copper concentration in the trunks of both Salix species, but had no effect on the levels of other heavy metals in the biomass. Better receptiveness of willow clones for mycorrhizal inoculum was accompanied by more efficient heavy metal extraction from contaminated soils. This furnishes a guideline for willow clone selection for phytoremediation. Increased biomass production with relatively constant metal concentrations constitutes an advantage of mycorrhizal formation in willows for phytoremediation of contaminated soils (Hrynkiewicz a d Baum, 2013).

MYCORRHIZAE DECREASE ALUMINIUM PHYTOTOXICITY IN ACIDIC SOILS

Elevated aluminium levels in soil exert an adverse influence on plant physiology and growth leading to poor soil productivity. When confronted with the stress of acidic soil, plants develop mechanisms that undermine the activity of aluminium and its toxicity, by exudation of Al-chelating compounds like organic acids into the rhizosphere or by storage of Al-organic acid complexes sequestered in the plant cells. Arbuscular mycorrhizal fungi augment resistance of the host plant to acidity and high levels of Al in the soil (Seguel et al., 2013).

Mycorrhizae Alter Plant Responses to Global Change

Kivlin et al. (2013) studied how host plants associated with arbuscular mycorrhizal fungi, ectomycorrhizal fungi, class I leaf endophytes, and dark septate endophytes reacted to global warming, enriched carbon dioxide, drought, and nitrogen deposition. In all global changes with the exception of high carbon dioxide concentration, the aforementioned fungal symbionts modified plant responses to the global changes. Fungal symbionts often elevated the plant biomass in response to global changes. However, the advantages of symbiosis were undermined by enhanced nitrogen deposition. Nitrogen deposition and drought elicited the most intense fungal mediation of plant responses. Thus considering plant-fungal symbiosis is important to predicting ecosystem response to global changes (Kivlin et al., 2013).

Mycorrhizae Enhance Plant Tolerance to Adverse Conditions

Prometryn and acetochlor are weedicides utilized in agriculture. Li et al. (2013) examined the effects of the two weedicides on spore germination, hyphal elongation, biomass and malondialdehyde content of carrot hairy roots by employing a strict *in vitro* cultivation system associating the Ri T-DyNA-transferred carrot hairy roots with the arbuscular mycorrhizal fungus *Glomus etunicatum.* Root colonization, daughter spore production and the abundance of hyphae possessing succinate dehydrogenase and alkaline phosphatase activities were also studied. Both prometryn and acetochlor produced adverse effects on the arbuscular mycorrhizal fungus. However, the arbuscular mycorrhizal symbiotic system improved biomass, minimized malondialdehyde accumulation and alleviated lipid peroxidation in carrot hairy roots and attenuated damage in host plants, thus augmenting plant tolerance to unfavorable conditions. The data on impact of prometryn and acetochlor on transformed carrot hairy roots/ arbuscular mycorrhizal fungus association under strict *in vitro* culture conditions may guide the application of the two weedicides in agriculture (Li et al., 2013).

Mycorrhizal Symbiosis Influences Host Plant Sulfur Starvation Response

The metabolite levels in *Medicago truncatula* leaves revealed that phosphate stress was more influential than sulfur stress on plant metabolism; there was no demand for sulfur during intense phosphate starvation. However, when phosphate nutrition was adequate, mycorrhizal colonization mitigated the sulfur stress responses due to symbiotic sulfur uptake, demonstrating the effect of mycorrhizal sulfur transfer on plant metabolism (Siehet al., 2013).

Plant Hormones in Arbuscular Mycorrhizal Symbiosis

Arbuscular mycorrhizal symbioses are important for nutrient acquisition in >80 % of terrestrial plants. Compared with wild-type plants, there was an appreciable increase in arbuscular mycorrhizal colonization of pea roots in gibberellin-deficient na-1 mutants. GA3 application reversed the phenomenon. Colonization was lessened in mutant la cry-s devoid of gibberellin signalling DELLA proteins. These changes were accompanied by altered expression of mycorrhizal colonization-associated genes. There were no changes in colonization.in the brassinosteroid-deficient lkb mutant. In pea roots, gibberellins inhibited arbuscule formation. DELLA proteins play an important role in this response within the root cells (Foo et al., 2013).

The plant cell development for intracellular accommodation of arbuscular mycorrhiza fungi is triggered by a signal transduction pathway that employs calcium spiking in the nucleus as the second messenger. Calcium spiking is stimulated by chitooligosaccharides produced by arbuscular mycorrhizal fungi probably perceived through LysM domain receptor kinases. Fungal infection and calcium spiking are coordinated in such a way that only cells hosting the fungus have high-frequency spiking. Delivery of minerals by arbuscular mycorrhizal fungi takes place at arbuscules, which are arborescent hyphal structures, in plant cortical cells. Nutrients are taken up at a plant-derived periarbuscular membrane that envelops fungal hyphae and has a specific transporter composition important for symbiotic efficiency. The vast majority of tested plant hormones are involved in the control of arbuscular mycorrhiza development (Gutjahr Parniske. 2013).

Conclusion

Mycorrhizae increase mineral and nutrient uptake by plants, and enhance the ability of host plants to withstand adverse environmental conditions like predators, toxins, drought and salinity. Mycorrhizae improve plant biomass production and growth, play a role in lateral root formation and phytomediation, modify physical and chemical properties of the soil, and alter plant responses to global change. Therefore mycorrhizal fungi contribute substantially in their symbiotic association with the host plants.

References

Averill C, Turner BL, Finzi AC. Mycorrhiza-mediated competition between plants and decomposers drives soil carbon storage. *Nature.* 2014; 505(7484): 543-5.

Barber NA, Kiers ET, Theis N, Hazzard RV, Adler LS. Linking agricultural practices, mycorrhizal fungi, and traits mediating plant-insect interactions. *Ecol. Appl.* 2013; 23(7):1519-30.

Duhamel M, Pel R, Ooms A, Bücking H, Jansa J, Ellers J, van Straalen NM, Wouda T, Vandenkoornhuyse P, Kiers ET. Do fungivores trigger the transfer of protective metabolites from host plants to arbuscular mycorrhizal hyphae? *Ecology.* 2013;94(9):2019-29.

Estrada B, Aroca R, Barea JM, Ruiz-Lozano JM. Native arbuscular mycorrhizal fungi isolated from a saline habitat improved maize antioxidant systems and plant tolerance to salinity. *Plant Sci.* 2013; 201-202:42-51.

Foo E, Ross JJ, Jones WT, Reid JB. Plant hormones in arbuscular mycorrhizal symbioses: an emerging role for gibberellins. *Ann. Bot.* 2013; 111(5):769-79.

Fusconi A. Regulation of root morphogenesis in arbuscular mycorrhizae: what role do fungal exudates, phosphate, sugars and hormones play in lateral root formation? *Ann. Bot.* 2014;113(1):19-33.

Gutjahr C, Parniske M. Cell and developmental biology of arbuscular mycorrhiza symbiosis. *Annu. Rev. Cell Dev. Biol.* 2013; 29:593-617.

Hrynkiewicz K, Baum C. Selection of ectomycorrhizal willow genotype in phytoextraction of heavy metals. *Environ. Technol.* 2013; 34(1-4):225-30.

Kivlin SN, Emery SM, Rudgers JA. Fungal symbionts alter plant responses to global change. *Am. J. Bot.* 2013; 100(7):1445-57.

Leung HM, Leung AO, Ye ZH, Cheung KC, Yung KK. Mixed arbuscular mycorrhizal(AM) fungal application to improve growth and arsenic accumulation of Pterisvittata (As hyperaccumulator) grown in As-contaminated soil. *Chemosphere.* 2013; 92(10):1367-74.

Li X, Miao W, Gong C, Jiang H, Ma W, Zhu S. Effects of prometryn and acetochlor on arbuscular mycorrhizal fungi and symbiotic system. *Lett. Appl. Microbiol.* 2013; 57(2):122-8.

Li Y, Wang H, Wang W, Yang L, Zu Y. Ectomycorrhizal influence on particlesize, surface structure, mineral crystallinity, functional groups, and elemental composition of soil colloids from different soil origins. *Scientific World Journal.* 2013; 2013:698752.

Maya MA, Matsubara Y. Influence of arbuscular mycorrhiza on the growth and antioxidative activity in cyclamen under heat stress. *Mycorrhiza.* 2013; 23(5):381-90.

Orłowska E, Przybyłowicz W, Orlowski D, Mongwaketsi NP, Turnau K,Mesjasz-Przybyłowicz J. Mycorrhizal colonization affects the elemental distribution in roots of Ni-hyperaccumulator Berkheya coddii Roessler. *Environ. Pollut.* 2013; 175:100-9.

Pangesti N, Pineda A, Pieterse CM, Dicke M, van Loon JJ. Two-way plantmediated interactions between root-associated microbes and insects: from ecology to mechanisms. *Front. Plant Sci.* 2013; 4:414.

Pena R, Tejedor J, Zeller B, Dannenmann M, Polle A. Interspecific temporal and spatial differences in the acquisition of litter-derived nitrogen by ectomycorrhizal fungal assemblages. *New Phytol.* 2013; 199(2):520-8.

Phieler R, Voit A, Kothe E. Microbially supported phytoremediation of heavy metal contaminated soils: strategies and applications. *Adv. Biochem. Eng. Biotechnol.* 2013 May 30. [Epub ahead of print].

Rineau F, Shah F, Smits MM, Persson P, Johansson T, Carleer R, Troein C,Tunlid A. Carbon availability triggers the decomposition of plant litter and assimilation of nitrogen by an ectomycorrhizal fungus. *ISME J.* 2013; 7(10):2010-22.

Roger A, Colard A, Angelard C, Sanders IR. Relatedness among arbuscularmycorrhizal fungi drives plant growth and intraspecific fungal coexistence. *ISME J.* 2013;7(11):2137-46.

Seguel A, Cumming JR, Klugh-Stewart K, Cornejo P, Borie F. The role of arbuscular mycorrhizas in decreasing aluminium phytotoxicity in acidic soils: a review. *Mycorrhiza.* 2013; 23(3):167-83.

Sieh D, Watanabe M, Devers EA, Brueckner F, Hoefgen R, Krajinski F. The arbuscular mycorrhizal symbiosis influences sulfur starvation responses of *Medicago truncatula. New Phytol.* 2013; 197(2):606-16.

Smithwick EA, Eissenstat DM, Lovett GM, Bowden RD, Rustad LE, Driscoll CT.Root stress and nitrogen deposition: consequences and research priorities. *New Phytol.* 2013; 197(3):712-9.

Tian M, Chen YL, Li M, Liu RJ. [Structure and function of arbuscular mycorrhiza: a review]. *Ying Yong Sheng Tai Xue Bao.* 2013; 24(8):2369-76.

Zeng Y, Guo LP, Chen BD, Hao ZP, Wang JY, Huang LQ, Yang G, Cui XM, Yang L,Wu ZX, Chen ML, Zhang Y. Arbuscular mycorrhizal symbiosis and active ingredients of medicinal plants: current research status and prospectives. *Mycorrhiza.* 2013; 23(4):253-65.

In: Mycorrhizas
Editor: Elaine Warwick
ISBN: 978-1-63321-436-1

Chapter 4

STUDY OF NITROGEN ASSIMILATION IN ECTOMYCORRHIZA BY RNAi-BASED GENE SILENCING: THE ROLE OF THE *LACCARIA BICOLOR* NITRATE TRANSPORTER

Minna J. Kemppainen and Alejandro G. Pardo*
Laboratorio de Micología Molecular, Departamento de Ciencia y Tecnología, Universidad Nacional de Quilmes, and Consejo Nacional de Investigaciones Científicas y Técnicas (CONICET). Bernal, Provincia de Buenos Aires, Argentina

ABSTRACT

In soil N is present in inorganic forms as ammonium (NH_4^+) and nitrate (NO_3^-), and as organic compounds such as amino acids, peptides and proteins. However, availability of N appears to be a limiting factor for forest tree growth in most temperate and boreal ecosystems where a significant pool of N is present in complex organic compounds.

Ectomycorrhizal fungi play a major role in soil N cycling by their utilization of both inorganic and organic forms. Nutrients, including N, are mobilized from soil by fungal mycelia and transferred to the host plant.

* Corresponding author: Alejandro G. Pardo. Phone: 54-11-4365-7100 (ext. 5639), fax: 54-11-4365-7132, e-mail: apardo@unq.edu.ar.

On the other hand, the host feeds the fungal symbiont with carbon compounds derived from photosynthesis. Despite the widely accepted importance of ectomycorrhizas on forest tree nutrition and N cycling, the control of basic N metabolism in ectomycorrhizal fungi is still poorly studied. Therefore, resolving how ectomycorrhizal fungi regulate their N adquisition will contribute to a better understanding of forest tree nutrition. *Laccaria bicolor* is an ectomycorrhizal basidiomycete which forms symbiosis with several boreal and temperate forest trees such as birch, pine and poplar.

Ectomycorrhizal fungi such as *L. bicolor* compete with saprotrophs and plants for the limited pool of nutrients. Ectomycorrhizal fungi are well known to utilize complex organic N sources especially abundant in acid boreal and temperate forest soils but mineral N is efficiently utilized as well. Ectomycorrhizal fungi appear to be well adapted to low mineral N concentrations and efficiently assimilate inorganic N as NH_4^+, NO_3^-, the first generally being the preferred N source. While NH_4^+ is the dominant mineral N form in undisturbed boreal and temperate forest soils the true availability of NO_3^- for ectomycorrhizal fungi in their habitat is not well established. The majority of studied ectomycorrhizal fungi shows a conserved capacity to grow on NO_3^- as the sole N source and some of them have even been reported to prefer NO_3^- over NH_4^+. This suggests that NO_3^- may play a more significant role as N source in forest ecosystems as traditionally believed. *Laccaria*, like many other ectomycorrhizal fungi, clearly prefers NH_4^+ but it also grows well on NO_3^-. Its genome encodes for one copy of each NO_3^- assimilation genes: the transporter *Lbnrt,* the NO_3^--reductase *Lbnr* and the NO_2^- -reductase *Lbnir*. The genes are organized in the so-called fungal high-affinity N-assimilation cluster (fHANT-AC). We found massive effects on the formation of ECM root tips of poplar when N-assimilation in *L. bicolor* was impaired by silencing of a selected NO_3^--metabolism gene, the nitrate reductase encoding gene *Lbnr*. Alteration of the symbiotic capacity was specific to the RNAi-induced reduction in N assimilation of the fungal partner. Silencing of *Lbnr* strongly impaired mycorrhiza formation on NO_3^- but not on NH_4^+. The participation of NO_3^- -utilization genes in the symbiotic interaction of *Laccaria* is hence clear.

We have extended the nitrate utilization RNAi-based gene silencing studies of the ECM basidiomycete *Laccaria bicolor* by altering the expression of *LbNrt*, the sole nitrate transporter encoding gene of the fungus. Here we report the first nutrient transporter mutants for mycorrhizal fungi. Silencing of *LbNrt* results in fungal strains with minimal detectable *LbNrt* transcript levels, significantly reduced growth capacity on nitrate and altered symbiotic interaction with poplar. Transporter silencing also produces co-downregulation of the whole fHANT-AC. Moreover, this effect on the nitrate utilization pathway seems to be independent of extracellular nitrate or the nitrogen status of the fungus. Our results show a novel and key nitrate uptake-independent regulatory role for a eukaryotic nitrate transporter. The possible cellular mechanisms underneath this regulation are discussed in the light of the current knowledge on NRT2-type nitrate transporters.

1. Introduction

Access to nitrogen is one of the fundamental growth limiting factors in terrestrial and aquatic ecosystems. While the majority of soil nitrogen is organic, nitrate (NO_3^-) is the major inorganic nitrogen source assimilated by many eukaryotes such as fungi and plants. In the fungal kingdom this capacity to utilize nitrate is limited to the Dikarya (Ascomycota and Basidiomycota), and to some members of the mucorales (Sarbhoy, 1965; Slot and Hibbett, 2007). Temperate and boreal forest soil fungi, equally to their relatives from agricultural soils, show a widespread capacity to grow on nitrate (Yamanaka, 1999; Nygren et al., 2008; Gorfer et al., 2011), This strongly suggest that nitrate plays a relevant role, not only in agricultural systems where its role as the most dominant inorganic nitrogen source is well established, but in forest ecosystems as well (Booth et al., 2005; Miller et al., 2007; Myrold and Ritchie Posavatz, 2007; Dechorgnat et al., 2011). Moreover, recent field studies on climate change effects have demonstrated that elevated CO_2 exposure and soil warming both induce increased nitrate concentration in temperate forest soils (Butler et al., 2012; Schleppi et al., 2012). These findings propose that the relevance of nitrate for forest nitrogen cycling is an increasing global trend.

Many of the nitrate utilizing fungi in boreal and temperate forests soils are symbiotic and engaged in mycorrhizal interactions with plants.

Ectomycorrhiza (ECM) is the dominant form of mycorrhiza of boreal and temperate tree species and this interaction is formed by a vast number of basidiomycete and a minor number of ascomycete species. In ECM the fungus supplies the host plant with different nutrients, of which nitrogen and phosphorus are the most important, while the fungus receives carbon as plant photosynthetic compounds (Smith and Read, 2008). While the majority of ECM fungi are reported to prefer ammonium as an inorganic nitrogen source the capacity to grow on nitrate is also widely distributed among them (Nygren et al., 2008) and some ECM species even prefer nitrate over ammonium (Scheromm et al., 1990; Aouadj et al., 2000). Most importantly, the ECM interaction can dominate the whole nitrogen uptake of the host tree which now occurs via the extraradical mycelium of the fungal partner (Gobert and

Plassard, 2002; 2007; Guescini et al., 2003; Bailly et al., 2007). This fungal dominance over the plant nutrient acquisition makes the uptake and control of nitrogen source utilization, such as nitrate, in ECM fungi of special interests as it can be expected to have a significant impact in nutrient acquisition of the symbiotic host plants and nutrient cycling in forest ecosystems.

Fungal utilization of nitrate requires its uptake by specific plasma membrane NRT2-type high-affinity nitrate transporters (NRT) and its sequential cytosolic reduction from nitrate to nitrite and further to ammonium by nitrate reductase (NR) and ferrodoxin-independent nitrite reductase (NIR) respectively. These nitrate metabolism genes are often present as a cluster, forming the so called fHANT-AC (fungal high affinity nitrate assimilation cluster). The fHANT-AC can be found in the majority of nitrate growing ascomycetes and basidiomycetes to which it was most likely horizontally transferred during the evolution from a lineage leading to oomycetes. While higher plants have multiple high- (HATS) and low- (LATS) affinity nitrate transporters (i.e. 53 NRT1- and 7 NRT2- type transporters in *Arabidopsis*; Dechorgnat et al., 2011), fungi show a very reduced nitrate transporter repertoire. Fungal transporters belong to the NRT2-type of nitrate transporters and the species studied harbor one or two transporter genes (Slot et al., 2007). Further differentiation to high- and low-affinity transporters and differential expression control has been demonstrated in some fungi which harbor the two transporter copies (Machín et al., 2000; Zhou et al., 2000; Unkles et al., 2001; Slot et al., 2010).

The fact that fungal nitrate growth relays on the activity of such a limited number of genes has made these organisms a target for intensive study of nitrate utilization in eukaryotes.

While the basic regulatory circuits in fungi have been identified more than 30 years ago, the precise molecular mechanisms responsible for control of fungal nitrate growth have been described only recently (Marzluf, 1997; Wong et al., 2008).

The current information on nitrate utilization gene regulation in fungi comes almost entirely from studies on saprotrophic and pathogenic ascomycete species such as *Aspergillus nidulans*, *Neurospora crassa* and *Magnaporthe grisea*, and on the ascomycete yeast *Hansenula polymorpha* (Crawford and Arst, 1993; Marzluf, 1997; Tao and Marzluf, 1999; Siverio, 2002; Wong et al., 2008; Rossi and Berardi, 2009).

Unfortunately, the information on genetic control of nitrate growth in other taxa, such as basidiomycetes, is minimal. This lack of information has special implications on ECM research.

Utilization of nitrate has a high energy cost and it is therefore shown to be under strict transcriptional feedback regulation in fungi, algae and plants (Crawford and Arst, 1993; Caddick et al., 1994; Berger et al., 2008; Fernandez and Galvan, 2008). The transcriptional activation of nitrate utilization genes, under sufficient carbon supply, requires in saprotrophic ascomycetes both the presence of nitrate as an inductor and the absence of a preferred nitrogen source, ammonium or L-glutamine as repressors (Marzluf 1997; Wong et al., 2008). The intracellular nitrate is believed to mediate this pathway induction response causing nuclear localization, and consequently DNA binding, of the nitrate-specific activating transcription factors (i.e. NirA in *A. nidulans*, Nit4 in *N. crassa*) (Fu et al., 1995; Punt et al., 1995). Promoter binding of these transcription factors depends on the synergistic action of global nitrogen status-sensing, positive acting GATA-binding factors (AreA, *A. nidulans*; Nit 2, *N. crassa*) (Kudla et al., 1990; Fu and Marzluf, 1987; Scazzocchio, 2000; Bernreiter et al., 2007). These GATA-factors mediate nitrogen metabolite repression (NMR) of nitrate and other secondary nitrogen utilization pathways in filamentous ascomycetes in the presence of preferred nitrogen sources such as ammonium and L-glutamine. The nuclear localization and DNA binding of both the nitrate-specific activating factor of NirA/Nit4-type and the positive acting GATA-factor is required for full transcriptional activation of nitrate utilization genes (Narendja et al., 2002; Mo and Marzluf, 2003; Muro-Pastor et al., 2004; Berger et al., 2006; 2008). However, AreA-responsive genes can also became activated by nitrogen starvation as lack of nitrogen has been demonstrated to cause gradual accumulation of the GATA-factor in nucleus and leads to Nit4/NirA-type inductor-independent activation of the nitrate pathway (Todd et al., 2005; Berger et al., 2008). In saprotrophic ascomycetes this nitrogen starvation activation of nitrate utilization genes is however moderate when compared to fully induced nitrate growth conditions. The repressing effect of ammonium in NMR has been firmly linked to elevated cellular L-glutamine concentration in *A. nidulans* indicating that this molecule, and not ammonium, would be the real intracellular repressor of the nitrate metabolism pathway (Berger et al., 2008).

In the last few years genome sequencing has resulted in major advances in ECM research, allowing interspecies genome comparisons and whole genome expression analysis (Martin et al., 2008; 2010; Plett and Martin, 2011). Access to genomic sequences had also made possible to assess the full nitrogen transport gene repertoire of these symbionts (Lucic et al., 2008).

Despite these advances the understanding of nitrate utilization regulation in symbiotic ECM fungi is still very limited.

Furthermore, the information originating from saprotrophic ascomycetes is apparently not directly applicable to these symbiotic fungi. This has become strikingly clear as the expression analyses of nitrate utilization genes in the few ECM fungi studied this far (i.e. ascomycete *Tuber borchii* and basidiomycetes *Hebeloma cylindrosporum* and *Laccaria bicolor*) have revealed some unique regulation characteristics clearly separating the mycorrhizal lifestyle from saprotrofism (Jargeat et al., 2000; 2003; Montanini et al., 2002; 2006a;b; Rékangalt et al., 2009; Kemppainen et al., 2010). The most outstanding about the nitrate gene regulation in these ECM fungi is that they share a nitrate-independent induction pattern, which can affect the whole (*H. cylindrosporum* and *L. bicolor*) or part (*T. borchii*) of the nitrate metabolism pathway. Since a similar relaxed transcriptional control is detected in taxonomically distant basidio- and ascomycete ECM species this suggests that such regulation mode affecting the capture of inorganic nitrogen from soil could be fundamental for efficient function of the ECM symbiosis or could have played a role in the evolution of this interaction (Kemppainen et al., 2010).

The fungal nitrate utilization genes have also been shown to be upregulated in the ECM interaction (Guescini et al., 2003; 2007; Bailly et al., 2007; Rékangalt et al., 2009) and uncompromised fungal nitrate metabolism has been related to successful establishment of the ECM interaction between *L. bicolor* and poplar (Kemppainen et al., 2009). ECM has also been shown to cause downregulation of plant nitrate utilization genes which reveals the dominance of the fungal nitrate metabolism pathway over the host (Bailly et al., 2007). All these data support the relevance of the nitrate utilization pathway for functioning of the ECM interaction and consequently for nutrient cycling in temperate and boreal forest soils.

Laccaria is susceptible to *Agrobacterium* mediated transformation (AMT) and RNA silencing making it the first ECM fungal model whose gene expression can efficiently be manipulated (Kemppainen et al., 2005; 2009; Plett et al., 2011). The fungus can utilize nitrate and its nitrate assimilation genes consist of one nitrate transporter, (*LbNrt*), nitrate reductase (*LbNr*) and nitrite reductase (*LbNir*) which are present as a fHANT-AC.

Our previous studies on the expression pattern of this nitrate gene cluster have already revealed some of its extraordinary regulatory features (Kemppainen et al., 2010).

While the whole cluster is clearly under NMR in presence of ammonium, also relative strong basal transcript levels can be detected under repressive conditions suggesting a simultaneous capacity to utilize both ammonium and nitrate.

While transcript accumulation of the cluster genes can be observed under nitrate feeding, most striking is however the strong nitrate-independent activation of the whole cluster detected under nitrogen limitation (starvation) or when the fungus is growing on organic nitrogen. One of these non-repressive organic N sources is L-glutamine, molecule which in ascomycete systems has been considered responsible for NMR (Marzluf, 1997; Berger et al., 2008; Wong et al., 2008). The *Laccaria* fHANT-AC regulation seems perfectly optimized for maximal nitrate uptake in soils with spatial and temporal gradients of this anion and rich in organic N. This high energy cost regulation pattern of *Laccaria* fits perfectly with its habitat, which is the upper temperate and boreal forest soil horizon rich in organic nitrogen, and its lifestyle as symbiont obtaining carbon from the ECM host plant.

These postulations are further supported by recent data from *H. cylindrosporum* nitrogen utilization gene regulation on different C/N ratios and nitrogen sources (Avolio et al., 2012).

The simultaneous capacity of *Laccaria* to utilize nitrate and organic nitrogen, despite the carbon cost, thus appears to be a general feature of ectomycorrhizal basidiomycetes and concords with the proposed high nitrogen sequestration requirements of the symbiotic lifestyle.

The potential of RNA silencing in resolving regulatory circuits of *Laccaria* nitrate metabolism and ECM interaction was previously demonstrated by altering the expression of the nitrate reductase gene (*LbNr*). These studies provided the first genetic confirmation for the importance of efficient fungal nitrogen metabolism in establishment of the ECM interaction (Kemppainen et al., 2009). They also demonstrated that *LbNr* plays a minimal role in regulation of other *Laccaria* fHANT-AC genes as reported in some other fungi (Kemppainen et al., 2010). Here we have extended the RNAi-based studies of *Laccaria* fHANT-AC activity and control by targeting the nitrate transporter gene (*LbNrt*). The current knowledge of the regulatory role of fungal nitrate transporters in nitrate utilization originates purely from saprotrofic ascomycete species which show strict NMR and nitrate-dependent transcriptional gene activation. As the transcriptional activation of nitrate utilization genes in these fungi is shown to depend on the nitrate-specific activator, and as the nuclear activity of this activator requires nitrate within the cells, a fundamental role of the nitrate transporter in initiating the intracellular nitrate signaling is expected. On the other hand, *Laccaria* offers a unique opportunity to study the role of fungal a nitrate transporter in a symbiotic basidiomycete which shows very different, highly relaxed and apparently a totally nitrate-independent fHANT-AC regulation pattern.

RNAi provides some great benefits for this goal as the effects of severely altered transporter expression can be studied without completely abolishing gene activity and thus nitrate uptake. This allows exploring the regulatory effects of a fungal nitrate transporter under conditions which for null-mutants would be lethal, and the possible nitrate-independent and nitrate-dependent regulatory effects could not be distinguishable due to total lack of nitrate uptake. Here we describe the structural characteristics of *LbNrt* and its gene product, and report the phenotype of the first RNAi-transporter mutants in mycorrhizal fungi. Our results demonstrate the high efficiency of RNAi for studying regulatory circuits in ECM and reconfirm the fundamental role of efficient fungal nitrogen metabolism in establishment of the symbiosis. RNA silencing of *LbNrt* reveals a central role of its gene product in upregulation of the rest of the fHANT-AC. Such a nitrate-independent regulatory role has not been described in eukaryotic nitrate transporters before.

2. LbNrt DNA Sequence Analysis

Laccaria bicolor nitrate transporter gene *LbNrt* (JGI protein ID 254042) forms part of a fHANT-AC in the genome of the fungus (Figure 1A). The *LbNrt* and the nitrate transporter gene *LbNr* (protein ID 254066) are closely clustered, and separated only by 593 bp, but the nitrite reductase gene *LbNir* (protein ID 291348) is located almost 14 Kb upstream. Such an open fHANT-AC is not present in other studied nitrate utilizing basidiomycetes and this long *LbNrt* and *LbNir* separating sequence harbors three putative ORFs.

It also carries retroviral sequence relics, which suggest that the structure of *L. bicolor* fHANT-AC could have originated through a transpositional event (Kemppainen et al., 2010).

The predicted genomic coding sequence of *LbNrt* stretches over 1902 bp and it consists of eight exonic and seven intronic sequences (Figure 1B). The promoter sequence prediction with NNPP (version 2.2) identifies a putative minimal promoter (score 0.93) at 21-70 nucleotides upstream of the *LbNrt* coding sequence with the proposed transcription start site (TSS) at 30 nt upstream of the start codon. The analysis of the *LbNrt* 5′-flanking region (700 bp upstream from ATG) for transcription factor binding sites with TESS search reveals only three sequence motifs with similarity to DNA binding sites of NIT2/AreA-type global positive acting nitrogen regulator GATA-factor from *Neurospora crassa* and *Aspergillus nidulans.* The nuclear localization and multiple promoter-binding of this type of transcription factors are known to be the absolute requirement for activation of different nitrogen metabolism genes which are controlled via NMR in ascomycetes. The upstream region of *N. crassa* nitrate transporter *nit-10* is reported to carry altogether 23 possible NIT2 binding motifs (Gao-Rubinelli and Marzluf, 2004) and the nitrate transporter gene from the ectomycorrhizal ascomycete *Tuber borchii* (*TbNrt*) possess 21 putative GATA-factor binding sites (Montanini et al., 2006). Gene transcription studies have demonstrated that nitrate utilization by basidio-mycete fungi is under global ammonium repression. This resembles the NIT2/ AreA-dependent regulation mode of ascomycete fungi and searches of different basidiomycete genomic sequences generally result in detection of putative basidiomycete NIT2/AreA orthologues. The transcription of *Laccaria* fHANT-AC genes is also controlled via dominant ammonium signaled repre-ssion (Kemppainen et al., 2010) and *Laccaria* genome encodes for at least one putative AreA orthologue (protein ID 488576).

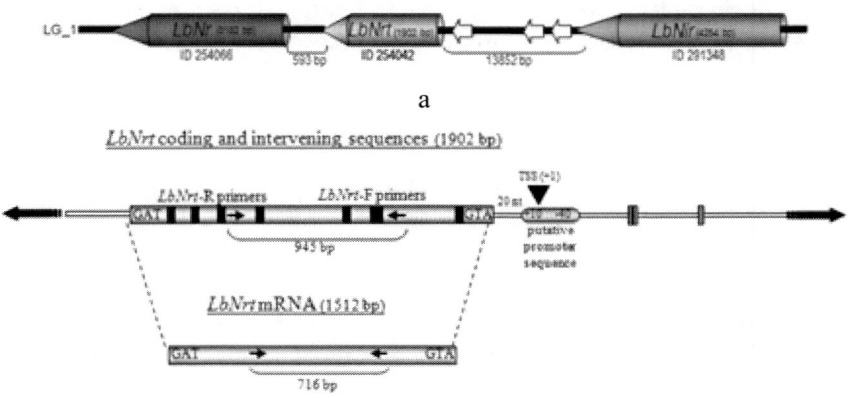

a

b

Figure 1. (A) Schematic representation of *L. bicolor* fHANT-AC structure. *LbNr*, nitrate reductase; *LbNrt,* nitrate transporter; *LbNir*, nitrite reductase genes. White arrows between the *LbNrt* and *LbNir* refer to detected ORFs separating the fHANT-AC genes. ID numbers presented are *L. bicolor* genome v2.0 assembly JGI protein ID numbers. (B) A schematic representation of *LbNrt* 1902 bp coding sequence and its upstream region. The black bars represent the location of seven introns and the yellow bar the minimal promoter detected with NNPP version 2.2. The putative transcription start site, TSS (+1), is marked with an arrow. The red bars in the *LbNrt* upstream sequence represent sites with similarity to NIT2/AreA-type global positive acting GATA-factor binding sites detected by TESS search. The binding sites of primers used for constructing the ihpRNA-expression silencing vector for *LbNrt* (*LbNrt*-R/*LbNrt*-F), and the amplicons sizes produced both from gDNA and mRNA are also indicated. For primer sequences see Kemppainen & Pardo (2013).

Moreover, this precise gene appears to be participating in control of nitrate utilization in the fungus (Kemppainen and Pardo, unpublished results).

However, not a single positive acting global nitrogen regulator basidiomycete GATA-factor has been characterized yet and their DNA binding motif characteristics are thus unknown.

The low number of detected nitrogen metabolism-related regulatory elements in the upstream sequence of *LbNrt* may partly reflect the possible different sequence requirements of basidiomycete specific transcription factors. Therefore, it is noteworthy that the *LbNrt* upstream sequence (700 bp) harbors altogether 11 core GATA sequences (GATA or TATC), several of which might be involved in nitrogen regulation of the gene even though their flanking nucleotides do not fulfill the NIT2 consensus sequence.

Besides the low number of putative positive acting GATA-factor -binding sequences neither did TESS search detected any ascomycete nitrate induction specific regulator of the NIT4-type binding sites in the upstream region of *LbNrt*. The promoter binding of NIT4 together with NIT2/AreA-type factors is known to be indispensable for strong activation of fHANT-AC genes when ascomycete fungi are exposed to nitrate. As with the global positive acting GATA-factor, this lack of NIT4-type binding site recognition might be related to unknown binding motifs characteristic of basidiomycete specific regulators. However, in the case of *Laccaria* this may also reflect the specific regulation mode of its fHANT-AC genes as the activity is demonstrated to be nitrate induction-independent (Kemppainen et al., 2010).

The same independency of nitrate has been shown for activity of fHANT-AC genes of other ectomycorrhizal basidiomycete, *Hebeloma cylindrosporum* (Jergeat et al., 2000; 2003; Avolio et al., 2012), and for the nitrate transporter (*TbNrt2*) in the ectomycorrhizal ascomycete *Tuber borchii* (Montanini et al., 2006). This lack of nitrate induction is an interesting feature as it is detected in phylogenetically distant species which share the same functional niche, the ectomycorrhizal symbiosis. Such a regulation mode thus might represent an important evolutive adaptation to the symbiotic lifestyle.

3. LbNrt Protein Sequence Analysis

The *LbNrt* encodes a 503 amino acid protein with an estimated molecular mass of 54.16 KDa. LbNRT shares the characteristics of the NNP (nitrate-nitrite porter) family of NO_3^- and NO_2^- porters (TC 2.A.1.8), members of which are found both in prokaryotes and eukaryotes.

The NNPs belong to the Major Facilitator Superfamily (MFS) proteins (TC 2.A.1) which are membrane transporters involved in symport, antiport and uniport of various substrates (Pao et al., 1998).

The nitrate transport via NNPs is shown to be energy demanding proton-coupled symport (McClure et al., 1990a;b; Miller et al., 2007) and the nitrate transporters of the NNP family in algae, higher plants and fungi are referred as NRT2 (Forde, 2000; Galván and Fernández, 2001).

The sequence analysis of LbNRT and a set of fungal and plant NRT2 proteins together with bacterial NNP sequences places LbNRT within a strongly supported clade of fungal NRT2 together with other basidiomycete nitrate transporters while the ascomycete transporters form their own clade (Figure 2).

Some ascomycete fungi, such as *Aspergillus* species, harbor several NRT2 isoforms and their sequence similarity indicates that these have originated from gene duplication events. This NRT2 duplication is apparently a sporadic event in fungi and also differential gene losses have happened within fungal lineages.

A second NRT2 gene paralogue can be found in some mushroom forming basidiomycete fungi such as *Hebeloma edurum,* but this duplication was probably later lost in the ECM fungus *H. cylindrosporum*. Interestingly, a single nitrate transporter is not a fully dominant feature of *Laccaria* genus either as one *Laccaria* species is demonstrated to have undergone a recent NRT2 gene duplication (Slot et al., 2010).

An important exception among the analyzed fungal protein sequences is the NRT2 of the ascomycete *Trichoderma reesei*. This transporter clusters together with the basidiomycete proteins and a horizontal gene transfer from basidiomycetes has been proposed to be behind the nitrate utilization capacity of this ascomycete species (Slot et al., 2007).

The typical membrane topology of MFS proteins consists of sets of six transmembrane helices which are connected by a cytosolic loop. This structure suggests that the early evolution of MFS proteins has probably involved a duplication of the transmembrane domains (Pao et al., 1998). The NRT2 transporters share the general 12 helical MFS structure and the fungal NRT2 proteins (TC 2.A.1.8.5) belong to the NNP subfamily II. It is characterized by relative short N- and C-terminal sequences outside the 12 transmembrane helices, a long hydrophilic central loop separating the 6th and the 7th helices with a conserved protein kinase C recognition motif S/T-x-R/K (Forde, 2000). The secondary structure topology models of the LbNRT were created by using three different prediction methods (TMHMM2, TopPred and TMpred).

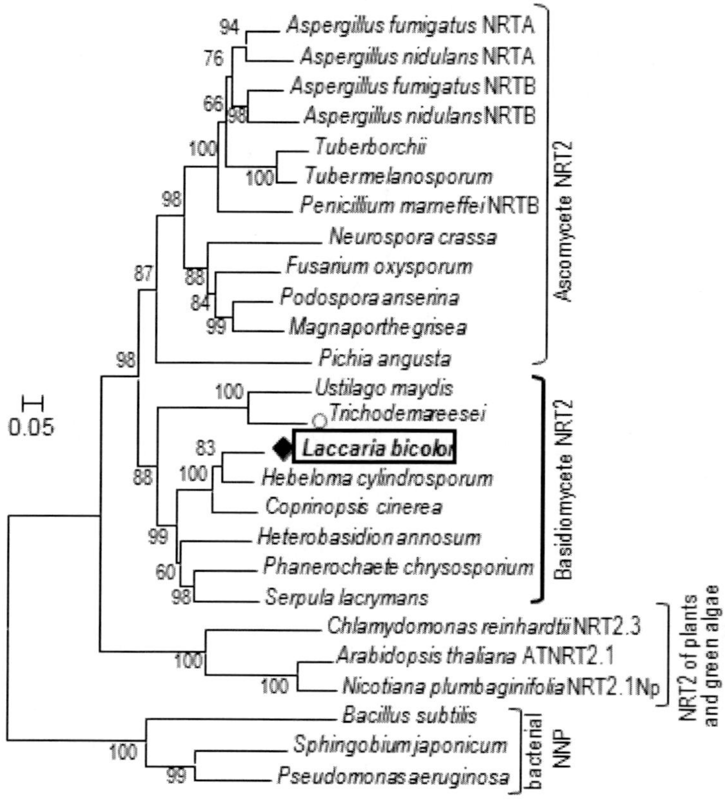

Figure 2. Phylogenetic analysis of *Laccaria* nitrate transporter with a set of fungal, algal and higher plant NRT2 and bacterial NNP proteins. The aligning of the predicted polypeptides (ClustalW), the Neighbor-Joining analysis (using Poisson correction and 1000 bootstrap replicas) and the visualization of the generated phylogenetic trees was performed with the MEGA 4.0 software package. For sequence sources see Kemppainen & Pardo (2013).

All three methods used proposed for LbNRT a conserved fungal-type NRT2 structure of 12 transmembrane helices, separated by an extended cytosolic sequence loop (71-72 aa, depending on the prediction program used), extended cytosolic N-terminal sequence (44-47 aa) and a very short cytosolic C-terminal end (0-3 aa) (Figure 3A). The analysis of the N-terminal sequence with SignalP-HMM prediction also suggests that this part of the protein could function as a signal anchor (probability score 0.853) and participate in association of the protein with the cell membrane via post-translational modifications.

A

	I	II	III	IV	V	VI	VII	VIII	IX	X	XI	XII
TMHMM2	45-67	77-99	112-131	136-158	171-193	208-230	302-324	339-361	384-404	419-441	454-476	480-502
TopPred	48-68	85-105	111-131	141-161	171-191	208-228	301-321	342-362	384-404	413-433	451-471	480-500
TMpred	48-67	88-105	113-130	136-161	164-188	210-230	301-320	346-370	384-404	415-441	452-471	480-503

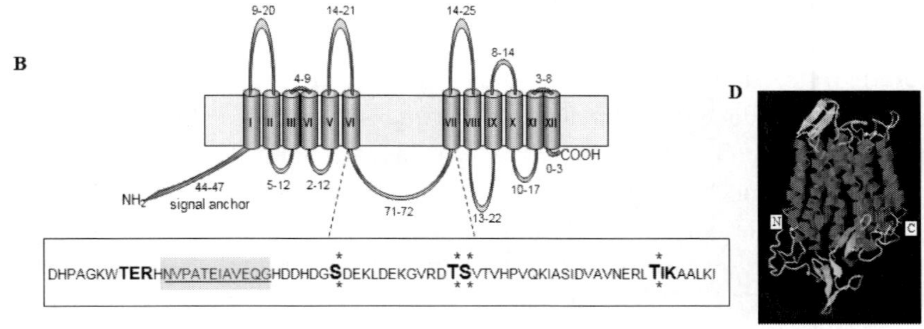

C

MFS I consensus sequence in NNPs: G-xxx-D-xx-G-x-R
MFS I in LbNRT: 100-**G** PLV **D** QY**G** P**R**-111
MFS II in LbNRT: 366-**G** YVG**D** L Ly rs -377

NNP I signature consensus sequence: F/Y/K-x-x-x-I/L/Q/R/K-x-G/A-x-V/A/S/K-x-G/A/S/N-L/I/V/F/Q-x1,2,-G-x-N/I/M-x-G-G/V/T/A
NNP I in LbNRT: 160-**F** D K N c V **G** T **A** N **A** **L** VGgw **G** M **M** Cp**G** **G**-183
NNP II in LbNRT: 449-N **G** Vm **S** - **G** **L** AGsf **G** T - L **G** **G**-465

Figure 3. (A) Secondary structure predictions of LbNRT by using TMHH2, TopPred and TMpred. The numbers (1-503) refer to amino acid positions and the roman numbers (I-XII) to predicted transmembrane α-helices. The numbers within the N- and C-terminal tails, and the trans-helical loops in the scheme of LbNRT refer to the range of amino acids proposed to be involved in each structure by the three secondary structure detection programs used. (B) Amino acid sequence of the long cytosolic loop between the VI and VII transmembrane helices and the putative posttranslational regulatory sequences in it. The amino acids in bold represent the conserved S/T-x-R/K protein kinase C recognition motives. The amino acids in bold with asterisks mark the potentially phosphorylated serine and threonine residues detected by NetPhos2. The shaded amino acids represent a weak PEST sequence detected by PESTfind. (C) MFS I, MFS II, NNPI and NNPII sequence signatures in LbNRT. The numbers refer to amino acids in the protein. The fully conserved amino acids of MFS and NNP signature sequences are marked in bold and underlined. The amino acids not present in the consensus sequence or the potentially additional ones are presented with small letters. The absence of an amino acid is marked with -. (D) A 3D model of LbNRT, amino acids 20-499, obtained via homology modeling based on *E. coli* glycerol-3-phosphate transporter crystal structure at the SWISS-MODEL server. N- and C-terminal ends of the protein are marked in the model. For details on methods see Kemppainen & Pardo (2013).

These secondary structure predictions obtained for LbNRT closely resemble the predicted membrane topology of NRT2 in another ectomycorrhizal basidiomycete, *H. cylindrosporum,* where a 48 aa N- terminal cytosolic sequence and a very short C-terminal tail (3 aa) has been proposed (Jargeat et al., 2003). The 3D model of the LbNRT was constructed via homology modeling using the crystal structure of the *E. coli* glycerol-3-phosphate trans-porter (PBD ID 1PW4; Huang et al., 2003) as a folding template. The model covering nearly the full length of LbNRT sequence (20-499 aa) was obtained and in this model the 12 transmembrane helices group to form a pore-like structure. Also the N-terminal tail, the long inter-helical loop and the C-terminus all face the same side, this being most probably the cytosolic side of the protein (Figure 3D). MFS proteins are known to carry repeated signature sequences (G-x-x-x-D/N-R/K-x-G-R-R/K), referred as MFS I and MFS II, and these are located between the transmembrane helices II and III, and VIII and IX, respectively. In NNPs this sequence motif can be found in the form G-x-x-x-D-x-x-G-x-R at the MFS I site while the MFS II is generally less conserved (Forde, 2000). In LbNRT the MFS I signature is detected at amino acid positions 101-110 between the helices II and III (GPLVDQYGPR). The less conserved MFS II motif carries only the conserved G and D residues (GYVGDLLYRS) and is located at position 367-376, between the transmembrane domains VIII and IV (Figure 3C).

The NPP protein family is also reported to carry another conserved signature motif within their transmembrane helix V (F/Y/K-x-x-x-I/L/Q/R/K-x-G/A-x-V/A/S/K-x-G/A/S/N-L/I/V/F/Q-x1,2,-G-x-N/I/M-x-G-G/V/T/A) and this signature is abbreviated as NNP I (Trueman et al., 1996; Pao et al., 1998).

Another related sequence, NNP II, is also generally repeated in the trans-membrane helix XI. The NNP I signature is present also in LbNRT in a slightly modified form between sites 161-182 (FDKNCVGTANALVGGWG MCPGG) and it runs from the cytosolic loop between the helices IV and V to the transmembrane part of the helix V (Figure 3C).

On the other hand, only a partial sequence related to the NNP II signature (GVMSGLAGSFGTLGG) could be detected within the predicted transmembrane helix XI of LbNRT (Figure 3C). Two conserved arginine residues present in the TM helices II and VIII of NRT2 transporters have shown to be needed for substrate binding and high affinity nitrate transport via NrtA in *A. nidulans* (Unkles et al., 2004).

Consistently with these results LbNRT also carries such conserved arginine residues in the given helices at amino acid positions 97 and 363.

4. Analysis of the Cytosolic Loop Sequence of LbNrt

The large cytosolic transhelical loop is a specific feature of fungal NRT2 proteins. This sequence typically carries protein kinase C recognition motifs S/T-x-R/K which suggests that the loop plays an important role in posttranslational regulation of the transporters via phosphorylations and dephosphorylations (Forde, 2000). Also LbNRT carries two conserved kinase C recognition motifs in its cytosolic loop sequence at sites 238-240 and 294-296 (Figure 3C). In addition to the protein kinase C motif, the analysis of the LbNRT sequence with the NetPhos 2. phosphorylation prediction program detected two potential serine (scores 0.988 and 0.992) and two threonine phosphorylation sites (scores 0.915 and 0.982) within the loop (Figure 3C).

Interestingly, the phosphorylation of serine residues in the cytosolic loop sequence of *Hansenula polymorpha* nitrate transporter YNT1 has recently been demonstrated to play a fundamental role in posttranslational control of this protein by sorting it to the plasma membrane in respect to the quality of nitrogen and availability of the carbon sources (Martín et al., 2011).

The presence of a potential protein kinase C recognition motif and Ser/Thr phosphorylation sites in LbNRT loop similarly to YNT1 strongly suggests that this part of the protein is important for post-translational regulation of nitrate transporter in *Laccaria* and in basidiomycetes in general.

The long cytosolic loop sequences of ascomycete NRT2s are also reported to contain regions resembling PEST sequences. The PEST sequences act as a signal for polyubiquitinylation which has been shown to initiate proteolytic cleavage and degradation. In *H. polymorpha* YNT1, its PEST sequence plays an important role in post-translational regulation of the protein.

The ubiquitination of the lysine residues in this sequence and consequent vacuolar degradation of YNT1 has been demonstrated to occur in response to glutamine in the growth medium (Navarro et al., 2006).

The PESTfind algorithm indicates the presence of a poor PEST sequence also within the loop of LbNRT (Figure 3C). Similar poor PEST sequences have been reported for NRT2 loop sequen-ces of several mushroom-forming basidiomycete species (Slot et al., 2010). The finding that *Laccaria* NRT2 together with other studied basidiomyce-tes carries a PEST region of such evolutive stability strongly suggests that a post-translational regulation via induced degradation could be a conserved feature between the ascomycete and basidiomycete NRT2 proteins.

5. Creation and Characterization of LbNrt-Silenced Strains of *Laccaria*

We wanted to study the *LbNrt* expression, its relevance for the symbiotic interaction and its possible role in *Laccaria* fHANT-AC regulation under variable nitrogen regimens via RNA silencing. Intron spacer hairpin RNA (ihpRNA) expression can efficiently launch gene silencing in *L. bicolor* (Kemppainen et al., 2009, Plett et al., 2011). *Agrobacterium*-mediated transformation in combination with the pHg/pSILBAγ –vector system (Kemppainen and Pardo, 2010) was used to create transgenic dikaryotic *Laccaria* strains carrying an ihpRNA expression cassette targeting *LbNrt*. This cassette consisted of 716 bp long inverted repeated sequence fragments of *LbNrt* cDNA (Figure 4A). Altogether 86 hygromycin B resistant strains were created with the pHg/pSγ/*LbNrt*Loop construct of which 43 randomly selected strains were evaluated for their growth on nitrate as N source in liquid medium. Of the screened strains 65 % showed severely reduced growth capacity on nitrate when compared to the wild type fungus. This growth phenotype was further confirmed for 10 randomly selected strains for their growth on solid nitrate medium with and without hygromycin B pressure. Independently of the presence of the selection antibiotic the affected strains had a thin superficial growth pattern and minimal hyphal penetration in the medium (Figure 4B). A similar growth phenotype has previously been linked to efficient *LbNr*-silencing in *Laccaria* (Kemppainen et al., 2009).

Two transformed strains (st. 2 and st. 4) with severely reduced capacity to grow on nitrate were used for further molecular analyses to demonstrate the affected expression status of the RNAi target, the presence of transgenes and to identify the T-DNA integration site in the genome the fungus. The initial semi-quantitative RT-PCR studies on nitrate grown strains confirmed that the *LbNrt* transcript levels in the tested strains were minimal in comparison to the wild type fungus (Figure 4C). The PCR testing for the presence of several sequence components of pHg/pSγ/*LbNrt*Loop T-DNA (Figure 4C) further demonstrated that both of the studied RNAi strains carried an intact *hph*-resistance and ihpRNA-expression cassettes. However, lack of amplification for the ampicilline resistance gene, which belongs to the T-DNA plasmid rescue component of pHg/pSILBAγ, in the strain 2 indicated that it carried a RB-truncated T-DNA integration.

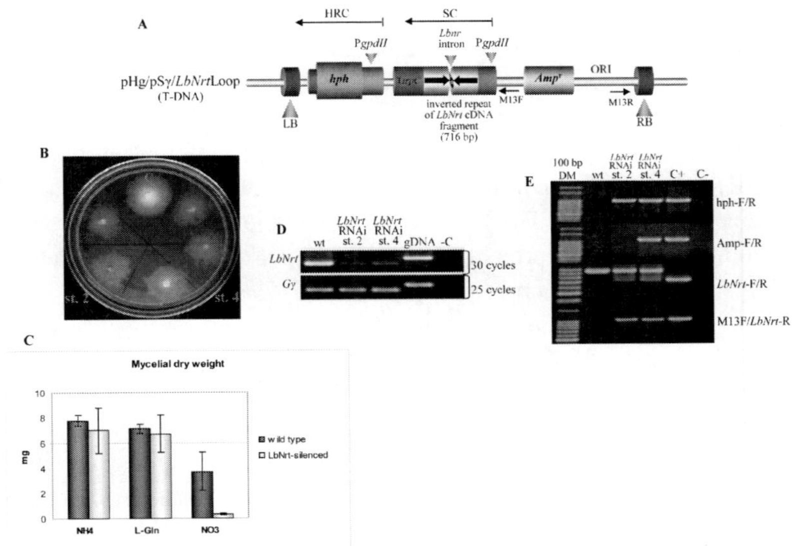

Figure 4. (A) Schematic representation of the transformation/silencing binary vector pHg/pSγ/*LbNrt*Loop T-DNA. P*gpdII*: *A. bisporus* glyceraldehyde-3-phosphate dehydrogenase promoter, *hph*: *E. coli* aminocyclitol phophotransferase gene which confers resistance to hygromycin B, T*trpC:* tryptophan synthetase terminator of *A. nidulans, Amp*^{r} : ampicilline resistance gene of *E. coli*, HRC: hygromycin B resistance cassette, SC: silencing cassette, LB, RB: T-DNA left border and right border repeats. M13R: M13/pUC-reverse sequencing primer site, M13F: M13/pUC sequencing primer site. (B) Growth of five pHg/pSγ/*LbNrt*Loop transformed *Laccaria* strains and a mock-transformant strain (C) on solid medium with 4 mM KNO_3 as nitrogen source. Strongly affected strains 2 and 4 used in further molecular analyses are marked. (C) The RT-PCR results on *LbNrt* expression with respect to heterotrimeric G-protein gamma subunit (JGI protein ID 181979) expression levels (Gγ) in the wild type fungus and in two pHg/pSγ/*LbNrt*Loop transformed, weak nitrate growth phenotype strains 2 and 4. gDNA: amplicons produced from genomic DNA template, -C: no template PCR control. (D) PCR detection of various pHg/pSγ/*LbNrt*Loop T-DNA elements in *LbNrt* RNAi strains 2 and 4. DM: DNA marker, C+: pHg/pSγ/*LbNrt*Loop as positive PCR control, C-: no template PCR control. *LbNrt*-F/R amplification shows both the gDNA and ihpRNA cDNA fragment amplification in the transformants but not in the wild type strain. M13F/*LbNrt*-R amplicon produced from the pHg/pSγ/*LbNrt*Loop T-DNA in RNAi strains (but not in the wild type strain) harbors both the ihpRNA cDNA arm and the *A.bisporus* gpdII promoter sequence. Lack of Amp^r amplification in the *LbNrt* RNAi strain 2 indicates a truncated T-DNA integration. Figure adapted from Kemppainen & Pardo (2013) where the methodology used in these experiments is detailed.

Consequently, the genomic T-DNA integration sites were recovered from the RNAi strain 4 by plasmid rescue and from the strain 2 by inverse PCR.

The results revealed that in both cases the T-DNA had integrated within ORFs with EST support.

Such a biased integration pattern in active genomic sites has been demonstrated before in *Laccaria* AMT under hygromycin B selection (Kemppainen et al., 2008). The RNAi strain 2 had the T-DNA integrated in a gene encoding a predicted protein of unknown function (ID 307960) and the strain 4 in a gene encoding a predicted protein (ID 479891).

Neither of these integration sites was the *LbNrt* locus, or physically close to genes of *Laccaria* fHANT-AC. This demonstrated that the altered *LbNrt* expression was not due to know nitrate utilization gene interruption and the ihpRNA-expression launched silencing was responsible for the minimal *LbNrt* transcript levels and consequently for the weak growth of the strains on nitrate.

In order to still confirm the connection between *LbNrt*-silencing and the reduced growth capacity specifically on nitrate, the biomass production of RNAi strain 2 was evaluated in liquid cultures under ammonium (*LbNrt* repressing N source), L-glutamine (non-repressive N source in *Laccaria*; Kemppainen et al., 2010) and nitrate feeding.

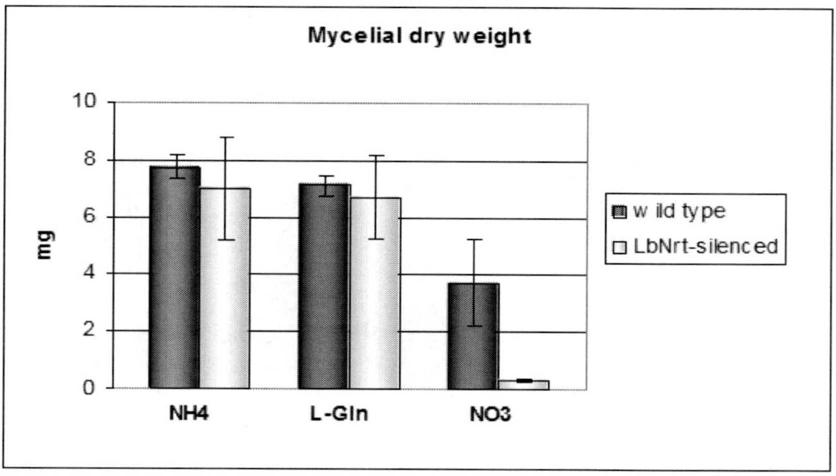

Figure 5. Biomass production of the wild type and the *LbNrt*-silenced strain 2 in liquid growth medium supplemented with ammonium (NH_4^+), L-glutamine (L-Gln) or nitrate (NO_3^-) as nitrogen source. For details on methods see Kemppainen & Pardo (2013).

The effect of *LbNrt*–silencing was clearly specific for nitrate growth as the capacity of the fungus to grow on ammonium or L-glutamine was unaltered but the biomass production on nitrate was less than 10 % of the wild type strain (Figure 5).

6. Silencing of LbNrt Affects the Symbiotic Interaction

Silencing of *Laccaria* nitrate reductase (*LbNr*) was previously shown to inhibit mycorrhization with poplar (Kemppainen et al., 2009). This result has pointed out the fundamental significance of uncompromised nitrogen metabolism of the fungal symbiont in formation of the ectomycorrhizal interaction. The two molecularly analyzed strongly *LbNrt*-silenced strains were equally tested for their symbiotic capacity with poplar.

The mycorrhization assay was run on growth media with minimal carbon availability for the fungus (10 g/l sucrose; a carbon source not utilizable by *Laccaria* as such) and combined nitrogen source (simultaneous high nitrate, 10 mM, and low L-glutamine 1.4 mM). This addition of the organic N source was used for compensating the weak biomass production of the silenced strains during co-cultivation with the plant. *Laccaria* fHANT-AC shows the unusual feature of not being repressed by L-glutamine and the fungus has capacity to utilize simultaneously both nitrate and an organic N sources (Kemppainen et al., 2010). The mycorrhization assay confirmed the results previously observed with *LbNr* silencing. While the wild type fungus vigorously interacted with the plant forming a symbiotic mantle covering the lateral roots of poplar (Figure 6A, panels 1-3), the interaction of *LbNrt*-silenced strains with the plant was minimal and reduced to loose hyphal growth on the surface of the roots (Figure 6B, panels 4-6).

The microscopic sections showed that no hyphal penetration between the epidermal and cortical cells has happened and consequently no Hartig net was formed by the *LbNrt*-silenced strains (Figure 6C, D). These results from *LbNrt*-silencing confirm the very low symbiotic capacity of *Laccaria* when its nitrogen metabolism is compromised and consequently no efficient nitrogen delivery from the fungus to the plant partner can be established.

7. Silencing of *LbNrt* Causes Nitrate-Independent Downregulation of *LbNr* and *LbNir*

Silencing of *LbNr* was previously shown not to affect the expression of other fHANT-AC genes (Kemppainen et al., 2010). This unaffected transcript status of *LbNrt* in a *LbNr*-silenced fungus was still confirmed by RT-PCR and Northern blotting (Figure 7) in two nitrate reductase silenced strains generated with the pHg/pSILBAγ-vector system (Kemppainen and Pardo, 2010).

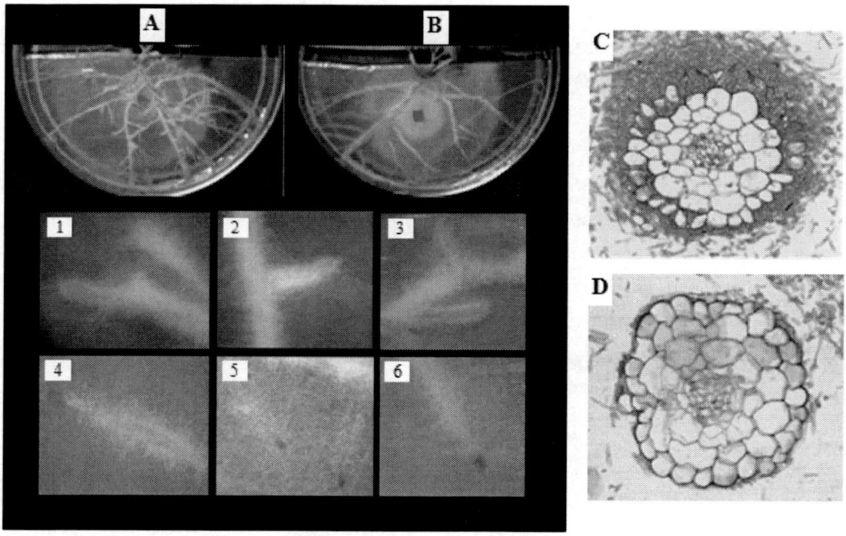

Figure 6. Summary of the mycorrhization assays with the wild type (A:1-3) and *LbNrt*-silenced strains (B: 4-6) of *Laccaria* with poplar. The wild type fungus efficiently formed the symbiotic mantle covering the lateral roots of poplar while the interaction between the plant and silenced strains was limited to loose mycelial coverage on some of the lateral roots. Microscopic sections of the symbiotic organs formed by the wild type fungus (C) had a fully developed Hartig-net but no such structure was observed in lateral roots of poplar confronted with the silenced *Laccaria* strains (D). Figure adapted from Kemppainen & Pardo (2013) where the methodology used in these experiments is detailed.

However, RT-PCR detection of lower amounts of *LbNr* and *LbNir* transcripts in nitrate grown *LbNrt*-silenced strain suggested that the silencing of the transporter causes a dramatic reduction of nitrate reductase transcripts and alteres the nitrite transporter expression as well (Figure 8A).

Identical semi-quantitative RT-PCR results were repeatedly obtained from three different pHg/pSγ/*LbNrt*Loop-transformed strains (st. 2, st. 4 and st. 13) which shared the strongly affected nitrate growth phenotype (Figure 9A). The relative fHANT-AC gene expression levels were calculated in two nitrate grown *LbNrt*-RNAi strains by quantitative real-time PCR (Figure 9B). Both of the studied strains had statistically very significant downregulation of both *LbNrt* and *LbNr* transcripts (0.2 - 0.5 % and 1.8 – 2.7 %, respectively) when compared to the wild type fungus. Also the *LbNir* transcript status was clearly affected in the RNAi strains but showed some variation in its degree.

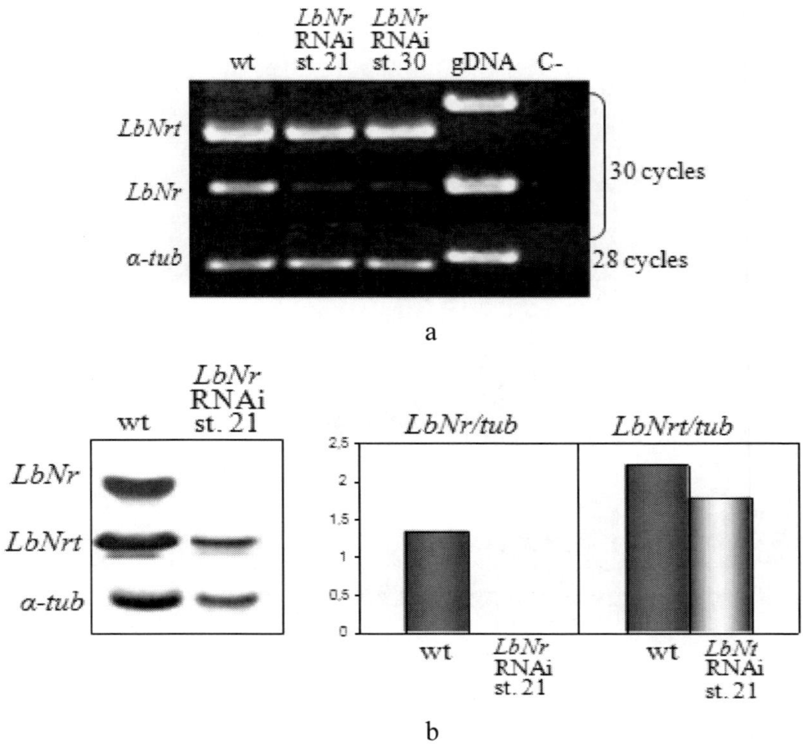

Figure 7. (A) RT-PCR assay showing the relative *LbNrt* and *LbNr* transcript levels of nitrate grown wild type and two strongly *LbNr*-silenced strains, 21 and 30 (for the description of the strains see Kemppainen and Pardo, 2010). gDNA: amplicons produced from genomic DNA template, C-: no template PCR control. (B) Northern blot detection and confirmation of the RT-PCR results for the unaffected *LbNrt* transcript status in the *LbNr*-silenced fungus. For details on methods see Kemppainen & Pardo (2013).

In the *LbNrt*-silenced strain 13, in which the residual *LbNrt* transcript level was also minor, the silencing co-downregulation effect on *LbNir* was statistically significant and the strain had only 31 % of the wild type expression level. On the other hand, in the slightly less *LbNrt*-silenced strain 2 the relative expression of the *LbNir* was of 60 %. This weaker silencing co-downregulation of *LbNir* compared to *LbNr* and its variability between the strains was already observed in semi-quantitative RT-PCR assays.

When correlated with the qPCR calculated *LbNrt*-silencing strength levels indications for a dose-effect can be observed, (i.e. more complete silencing of *LbNrt* initiates more profound co-downregulation of *LbNir* explaining the variation observed between the RNAi strains).

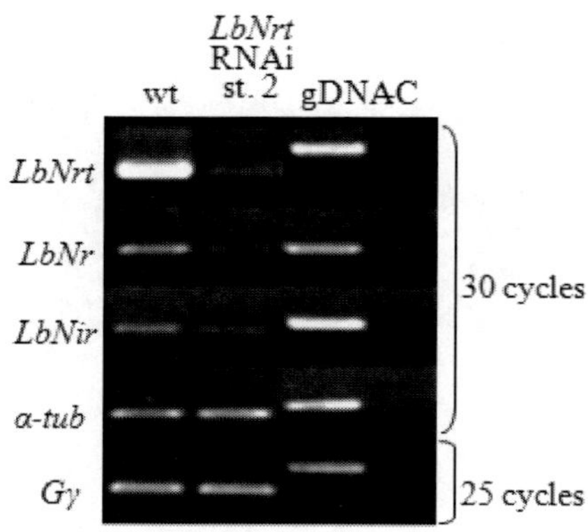

a

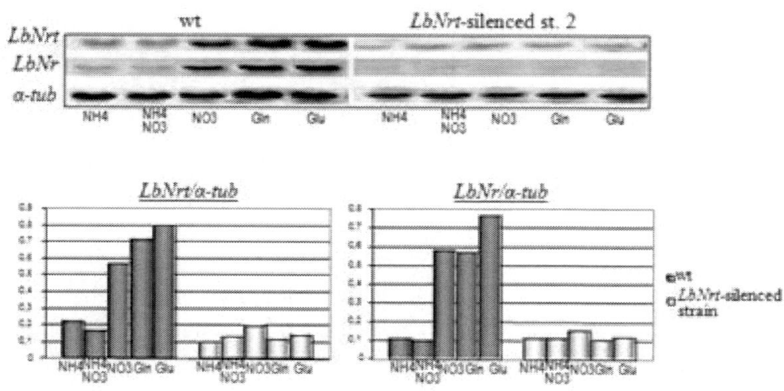

b

NH4: ammonium, NO3: nitrate, NH4NO3: simultaneous ammonium and nitrate feeding, Gln: L-glutamine, Glu: L-glutamate. Figure adapted from Kemppainen & Pardo (2013) where the methodology used in these experiments is detailed.

Figure 8. (A) RT-PCR assay showing the *LbNrt*, *LbNr* and *LbNir* transcript levels of nitrate grown wild type and strongly *LbNrt*-silenced strain 2 with respect to alpha tubulin (*α-tub*) and G-protein gamma subunit (Gγ) expression levels. gDNA: PCR amplicons produced from genomic DNA template, -C: no template PCR control. (B) Northern blot detection of the relative *LbNrt* and *LbNr* transcript levels in wild type and *LbNrt*-silenced strain 2 under variable nitrogen feeding conditions.

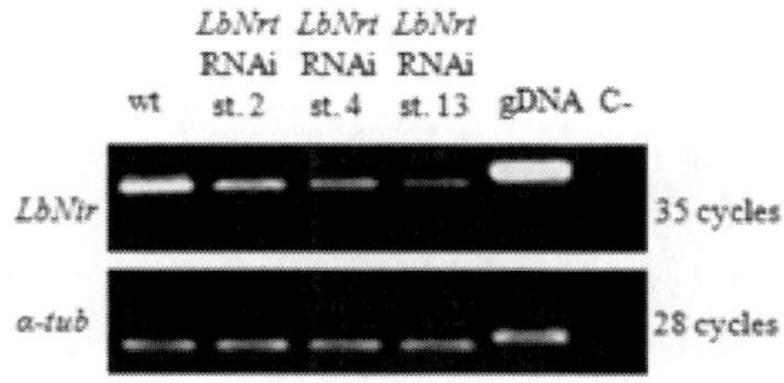

a

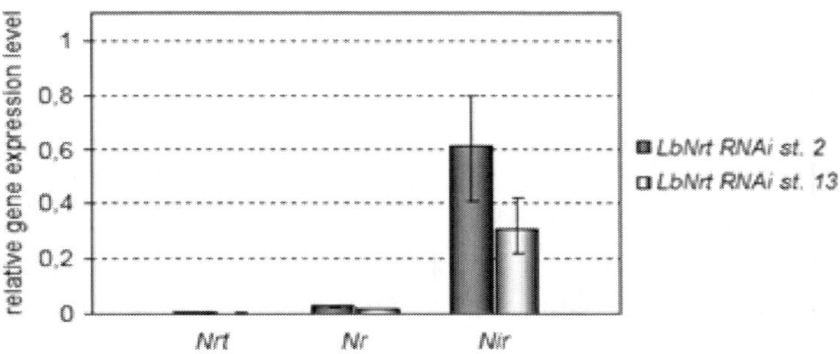

	LbNrt RNAi st. 2	*LbNrt* RNAi st. 13
LbNrt	0.005 P(H1)=0.003	0.002 P(H1)=0.000
LbNr	0.027 P(H1)=0.006	0.018 P(H1)=0.007
LbNir	0.608 P(H1)=0.07	0.31 P(H1)=0.003

b

Figure 9. (A) RT-PCR of *LbNir* transcript levels in nitrate grown wild type and strongly *LbNrt*-silenced strain 2, 4 and 13 in respect to alpha tubulin (*α-tub*) expression. gDNA: PCR amplicons produced from genomic DNA template, -C: no template PCR control. (B) Relative expression of fHANT-AC genes in two *LbNrt* silenced strains (2 and 13) in respect to nitrate grown wild type fungus determined by qPCR. The transcript levels were normalized with two *Laccaria* endogenous reference genes, β-actin (JGI protein ID 192701) and elongation factor 1 (ID 6073790). Hypothesis test P(H1)<0.05 indicates statistically significant difference between the sample and the control. Figure adapted from Kemppainen & Pardo (2013) where the methodology used in these experiments is detailed.

The semi-quantitative RT-PCR and qPCR results were further confirmed for *LbNrt* and *LbNr* by Northern blotting twice (Figure 8B).

Most importantly, the *LbNrt*-silencing linked co-downregulation of the nitrate utilization pathway was not detected only in presence of nitrate but clearly also in absence of it.

Minimal *LbNr* transcript levels were detected equally when the fungus was feeding on L-glutamine or L-glutamate even though these two organic N sources do not cause repression of *Laccaria* fHANT-AC. Identical results were obtained in nitrogen free growth conditions as well (data not shown). The Northern blot data finally revealed that the detected co-downregulatory effects of *LbNrt*-silencing on *LbNr* and *LbNir* could not be simply explained by reduced nitrate transporting activity of the *LbNrt*, and consequently by lower intracellular concentration of the substrate of the given metabolic pathway.

8. Discussion

Availability of nitrogen appears to be a tree growth limiting factor in most temperate and boreal ecosystems as in these soils significant amount of N accumulate in complex organic molecules, the majority of which are not directly available to plants or soil micro-organisms (Tamm 1991; Rennenberg et al., 2009). As ammonium forms the major inorganic nitrogen pool, especially in undisturbed forest soils, the importance of nitrate in boreal and temperate forests nitrogen cycling, has been considered minor (Blew and Parkinson, 1993; Jerabkova et al., 2006). However, the true availability of nitrate in these soils appears to be higher than traditionally believed as nitrate can show significant temporal and spatial concentration gradients (Stark and Hart, 1997; Leverman et al., 2000). This phenomenon is partly linked to the negative charge of clay which makes nitrate highly mobile in soil, and to soil micro-niche pH, temperature and oxygen status which influence the rate of nitrification (Strahm and Harrison, 2006). Due to the high mobility and patchy presence, soil micro-organism (algae, fungi, bacteria and archeae) and higher plants have developed efficient mechanisms to capture and utilize nitrate when encountered in their environment (Forde, 2000; Galván and Fernández, 2001; Cabello et al., 2004; Gojon et al., 2011; Luque-Almagro et al., 2011).

In accordance with the presence of nitrate in temperate and boreal forests soils, fungi from these ecosystems, many of which are mycorrhizal, show widespread capacity to utilize nitrate (Nygren et al., 2008). Since the ECM symbiosis can lead the fungal partner to dominate the soil nutrient uptake of the host tree the nitrate uptake capacity and control of ectomycorrhizal fungal species is of special interest as it can have forest ecosystem implications.

Our previous studies on nitrate utilization pathway genes, fHANT-AC, of the model ECM basidiomycete *Laccaria bicolor* have revealed its regulatory flexibility in comparison to saprotrophic ascomycetes (Kemppainen et al., 2010). The high activity of *Laccaria* nitrate usage genes under conditions which generally result repressive in saprotrophic fungi may well be linked to the symbiotic lifestyle of *Laccaria*. The efficiency to capture nitrogen from soil and transfer it to the ECM plant appears to be fundamental for the successful establishment of the symbiotic interaction with plant roots. This central role of efficient fungal nitrogen metabolism in ECM has received further support from severely reduced symbiotic capacity of *Laccaria* nitrate reductase RNAi mutants (Kemppainen et al., 2009).

Furthermore, a functional ECM interaction can offer the fungus a carbon source which is inaccessible to other soil micro-organisms. This potentially unlimited and steady energy niche may explain how the high carbon cost of the relaxed *Laccaria* fHANT-AC regulation could have evolved and resulted competitive over time.

We have continued the RNAi-based studies of *L. bicolor* fHANT-AC by silencing the nitrate transporter *LbNrt*. The *LbNrt* is the only nitrate transporter encoding gene of this *Laccaria* species and therefore indispensable for its growth on nitrate. The RNAi targeting of the *LbNrt* was accomplished via stable *Agrobacterium*-mediated transformation and expression of a partial transporter cDNA sequence as an intron-loop separated hairpin under a constitutive heterologous promoter (*A. bisporus* gpdII promoter).

RNA silencing in *Laccaria* has offered a novel way for studying ecto-mycorrhiza and metabolism in symbiotic fungi. Triggering of RNA silencing via ihpRNA expression using the pHg/pSILBAγ–vector system has previously been used for targeting *Laccaria* nitrate reductase gene (*LbNr*) and more recently for altering the expression of the mycorrhiza induced small secreted protein, MiSSP7 (Kemppainen et al., 2009; Plett et al., 2011).

Both of these studies have revealed important aspects of establishment and control of the EMC interaction and demonstrated the true value of RNAi as a tool for ECM research which has traditionally suffered from the fungal recalcitrance to genetic manipulation.

The herein presented *LbNrt*-silenced *Laccaria* strains are the first nutrient transporter mutants of ECM fungi to date. The silencing gene targeting was shown to be highly efficient and to reduce the *LbNrt* mRNA level under fHANT-AC gene expression inducing conditions to less than one percent of the wild type expression level. The gene silencing was also specific to nitrate metabolism as the biomass production on nitrate, but not on other nitrogen sources, such as ammonium and L-glutamine, was affected. The silencing-altered nitrate metabolisms status of *Laccaria* had also direct effect on its symbiotic capacity with poplar under predominant nitrate feeding conditions.

These results reconfirm the results obtained with *LbNr*-silenced strains (Kemppainen et al., 2009). The same lack of normal mantle development and Hartig-net formation was observed with *LbNrt*-silenced strains when growing on nitrate and confronted with the host tree.

Our previous studies demonstrated that the nitrate metabolism pathway in *Laccaria* shows a novel and apparently totally nitrate-independent activation pattern. Also the *LbNr*-silencing results proved that *Laccaria* nitrate reductase is not a key regulator of the two other nitrate pathway genes as wild type transcript levels of *LbNrt* and *LbNir* were detected in *LbNr*-silenced strains (Kemppainen et al., 2010, and results presented here). In the light of the previous data and especially considering the apparently nitrate-independent activation mode of *Laccaria* fHANT-AC genes, the detected co-downrecula-tion effect of *LbNrt*-silencing on *LbNr* and *LbNir* was unexpected. The reduced *LbNr* transcript levels were detected in *LbNrt*-silenced strains not only when the fungus was growing on nitrate but also in its absence.

As nitrate appears to have a dispensable role in control of fHANT-AC genes, why a strong, unaltered expression of the transporter would be required for further transcriptional activation of nitrate and nitrite reductase genes? Is the transporter-dependent influx of nitrate after all playing some role in *Laccaria* as fHANT-AC gene inducer? Can the detected co-downregulation be explained by the lack of high-affinity nitrate transporter-dependent cellular accumulation of nitrate traces, the so called *gratuitous induction hypothesis*, which has been postulated for the nitrate utilizing yeast *Hansenula polymorpa* and for the green algae *Chlamydomonas reinhardtii*?

Such intracellular accumulation of nitrate via the transporter activity on minimal concentration of nitrate in apparently nitrate free growth medium has been used as an explana-tion for the high-affinity nitrate transporter-dependent nitrate gene deregula-tion, which is observed in nitrate reductase lack-of-function mutants when these are growing without a nitrogen source (Rexach et al., 2002; Navarro et al., 2003).

Nitrate reductase mutants have been reported to show nitrate-pathway gene de-regulation (i.e. activation) also in several filamentous ascomycete fungi, especially in absence of exogenous nitrate (Unkles et al., 2001). However, this *gratuitous induction* –hypothesis, and thus "a hidden" role of transporter imported nitrate as *Laccaria* fHANT-AC inducer as well, seems an unlikely explanation for the nitrate transporter-dependent nitrate pathway co-downregulation which is detected under variable nitrogen growth conditions.

This is due to the fact that RNA silencing, even though demonstrated to lead to very significant reduction on the target transcript level, in contrast to null-mutants or lack-of-function mutants, does not lead to complete elimination of the target gene activity (as detected by RT-PCR, Northern blotting and qPCR for *LbNrt*). The basal *LbNrt* transcript levels were also detected under nitrate-feeding and in accordance with that, the *LbNrt*-silenced strains can grow on nitrate, even though this capacity is severly reduced (~10 % of wild type biomass). If intracellular nitrate, taken up by the transporter, would influence the activation of fHANT-AC genes one would expect that at least when growing on nitrate the residual *LbNrt* expression should be enough for carrying-out the nitrate uptake sufficient for induction of the rest of the nitrate pathway genes. This is not the case and neither does the silencing-linked co-downregulation appear to be a result of a possible nitrogen starvation response in *LbNrt*-silenced strains as the same fHANT-AC transcript reduction was observed when the mutants were growing on highly preferred, not fHANT-AC expression repressive nitrogen source, such as L-glutamine.

Could the data from *H. polymorpha* and *C. reinhardtii,* instead of the *gratuitous induction-hypothesis,* be interpreted as examples of nitrate transporter-linked, but exogenous nitrate uptake-independent, regulation of nitrate utilization pathway under nitrate starvation? In concordance with this interpretation the phosphorylation status of serine residues in the long cytosolic loop sequence of *H. polymorpha* NRT2 (YNT1) has been demonstrated to control the amount of the transporter at the plasma membrane and its degradation in vacuole. This phosphorylation is also required for rapid induction of other nitrate assimilation genes precisely under nitrogen-limiting conditions (Navarro et al., 2008, Martín et al., 2011). As the nitrate induction of *H. polymorpha* fHANT-AC genes has been demonstrated to depend on nitrate entering the cells via the given transporter (Navarro et al., 2003) the *gratuitous induction* was postulated. This would require increased membrane localization of the phosphorylated transporter protein whose transporter activity on nitrate traces would lead to increased cytosolic nitrate concentration and consequently to nitrate induction of other fHANT-AC genes (Navarro et al., 2008).

However, what is not clear is how this nitrogen-limiting growth condition in *H. polymorpha* differs from the fungal nitrogen starvation activation of nitrate utilization genes in general. In recent years, the lack of nitrogen source has been demonstrated to activate nitrate responsive genes in several ascomycetes and this activation is shown to depend only on the activity of positive acting GATA-factor and to be independent of nuclear localization of nitrate responding specific inducer-transcription factors (Berger et al., 2008). In the light of the current knowledge on nitrogen starvation activation and also on participation of intracellular produced nitrate (from nitric oxide) in activation of nitrate responsive genes in other ascomycetes (Berger et al., 2006; Schinko et al., 2010), the *gratuitous induction* hypothesis seems less probable to explain the nitrate transporter-dependent activation of nitrate metabolism pathway genes in *H. polymorpha* during nitrogen starvation as well. Therefore, both our result and results from *H. polymorpha* are pointing towards an exogenous nitrate-independent and direct role of the nitrate transporter protein itself in activation of other nitrate utilization genes. In the nitrate induction-independent system of *Laccaria* this happens under any fHANT-AC non-repressive growth condition (i.e. absence of ammonium) while in the nitrate induction responsive fungus *H. polymorpha* apparently only during nitrogen starvation. Interestingly, all of these conditions represent potentially sole positive nitrogen regulator GATA-factor dependent fungal gene activation modes. To our knowledge this is the first time that an altered fungal nitrate transporter expression has been linked to nitrate-independent regulation of other nitrate metabolism genes under non-nitrogen starvation growth conditions.

The information on basidiomycete nitrate assimilation and its control is minimal and to which extend data originating from ascomycete fungi is applicable to this group of fungi is not known. Therefore, how such nitrate independent and transporter-linked control of fHANT-AC genes is exploited in *Laccaria* can at this point be only speculative. LbNRT protein has a predicted structure of 12 transmembrane helices forming the membrane pore and a relative long cytosolic N-terminal tail with a putative signal anchor sequence. Similarly to other fungal nitrate transporters of the subfamily NRT2 II, LbNRT also carries a long cytosolic loop between the helices VI and VII. The presence of several post-translational regulation-liked sequence motives in this region strongly suggests that protein level regulation of LbNRT plays a role in *Laccaria* nitrogen utilization control. Posttranslational modifications have been shown to be fundamental in control of several eukaryotic nitrate transporters. Transporter phosporylation and ubiquitinylation has been demonstrated to play a central role in regulation of *H. polymorpha* YNT1 cellular localization (Navarro et al., 2006; Martín et al., 2011).

Also in *Aspergillus* NR-mutants the nitrate transporter mRNA and protein levels are significantly higher than in the wild type. Despite this elevated protein level the nitrate uptake is severely compromised and relative cell nitrate efflux is increased and the lack of uptake is probably controlled via posttranslational modification of the transporter protein (Wang et al., 2007). In *Arabidopsis* the PTR family nitrate transporter CHL1 is demonstrated to act as extracellular nitrate sensor and participate independently of its transporter activity in intracellular nitrate signaling. Besides the sensor function, the same protein can act, depending on nitrate conditions, as low- or high-affinity transporter. Both the sensor activity and modification of the substrate affinity are regulated via protein phosphorylations (Liu and Tsay, 2003; Ho et al., 2009). However, the PRT family nitrate transporters differ structurally from NRT2 transporters. Therefore, CHL1-type nitrate sensor and signaling mechanisms of the fungal transporters result unlikely.

Gene silencing can launch unwanted off-target effects on transcripts sharing homologous sequences. It has become evident that RNA silencing is not as sequence specific as previously believed and off-target effects can not be completely avoided, especially when long dsRNA-triggers are used. The characteristics of off-target triggering however vary between eukaryotes. In mammalian cells a 7 nt sequence complementary between the unwanted target and a siRNA (small interfering RNA) of a short hpRNA-trigger has been reported to launch off-target effects (Jackson et al., 2006). On the other hand, in *Drosophila* 16 nucleotides of contiguous homology are sufficient to attenuate gene activity (Haley et al., 2010). The minimal sequence homology requirements for initiating off-targeting in fungi are not well known yet and they also most probably vary between different fungal taxa. In *Magnaporthe oryzae* 13 nt perfect sequence homology was sufficient for simultaneous silencing of an unwanted target (Nguyen et al., 2008). We wanted to evaluate the possibility that the detected co-downregulation of *Laccaria* fHANT-AC genes in *LbNrt*-silenced strains could originate from such off-targeting. Homology region searches (Blastn, word size 7, expectation value 10) between the ihpRNA - trigger or the full length *LbNrt* transcript sequence and the *LbNr* transcript did not reveal longer than 11 nt exact match sequences between the two genes. Results of similar degree of homology were obtained with searches between the *LbNrt* and the control gene transcripts used in RT-PCR, Northern blotting and qPCR (α-tubulin, max. 10 nt matches; Gγ protein, 11 nt matches; β-actin, 12 nt matches; elongation factor 1, 12 nt matches). Also when three randomly selected and manually cured *Laccaria* gene transcripts (IDs 186476, 396609 and 469321) were compared against *LbNrt* similarly maximum 11 nt long perfect matches were detected, showing that this represents the degree of perfect homology stretches found between any *Laccaria* genes compared at random. Therefore, we propose that it is highly unlikely that the significantly reduced *LbNr* mRNA levels detected in various *LbNrt*-silenced strains would originate from off-target silencing.

The pSILBAγ-vector used for constructing the *LbNrt* silencing trigger expression cassette carries a *Laccaria LbNr* intronic sequence which forms the ihpRNA loop. For maximizing the intron-splicing 13 nt of genomic sequence before and after the intron were also incorporated in the vector. The detected co-downregulation arose an additional question about the possible role of these *LbNr* coding sequence nucleotides in the *LbNrt* ihpRNA-trigger as off-target initiators. The spacer sequences in hpRNA-triggers have been demonstrated not to act as silencing initiating component in maize and tobacco (Mitter et al., 2006; Houmard et al., 2007; Nishimura et al., 2009). However, in the nematode *Caenorhabditis elegans* dsRNA-loop sequence has been reported as capable of initiating RNA interference on homologous sequences via transitive silencing (Alder et al., 2003). The fate of the loop sequence of hpRNA-triggers in fungi is not clear. To assess whether pSILBAγ-launched silencing correlates with reduced *Laccaria* growth on nitrate and consequently with possible co-targeting effects on *LbNr* via the short homologous sequence present in the ihpRNA loop region, we tested the growth of ten *MissP7*-silenced and two mock transformation mutants strains (Plett et al., 2011) on solid nitrate agar medium. The growth of these strains was compared to the wild type fungus, using two strongly *LbNrt*-silenced and to two *LbNr*-silenced strains as controls for silencing altered nitrate growth in *Laccaria* (data not shown). Nine of the tested *MissP7*-silenced and the mock strains produced equivalent biomass growing on nitrate compared to the wild type. Only one of the *MissP7*-silenced strains produced weaker growth on nitrate but the growth of the strain was affected on ammonium as well. We take these results as confirmation that the short sequence in the ihpRNA-loop corresponding to *LbNr* cannot be hold responsible for the reduced *LbNr* and *LbNir* transcript levels in *LbNrt*-silenced *Laccaria* strains.

Our results raise several questions about the role of *LbNrt* in *Laccaria* fHANT-AC regulation. Our findings strongly suggest that the unaffected cellular level of the *LbNrt* gene product is a crucial regulator of transcriptional activation of rest of the nitrate utilization pathway in this basidiomycete. No similar regulatory role has been reported for any other eukaryotic nitrate transporter under nitrogen sufficient conditions this far. Could this indicate that besides being a transporter LbNrt also has a direct intracellular signaling function? How could such function be orchestrated? Any answers to this question can at this point be only hypothetical. However, increasing amount of experimental data is now pointing towards a nitrate uptake-independent sensor or signal transducer role of *Arabidopsis* major HATS protein, NRT2 family transporter NRT2.1 as well (Malamy and Ryans, 2001; Little et al., 2005; Remans et al., 2006; Gojon et al., 2011). This signaling role of the transporter, which affects lateral root development, acts under low nitrate conditions or in complete absence of nitrate varying its outcome in respect to carbon availability. The plasma membrane localization of this transporter and nitrate uptake requires also the partner protein NAR2.1 (Okamoto et al., 2006; Orsel et al., 2006). Both of these protein components are under posttranslational regulation even though the precise nature of it is not yet clear (Laugier et al., 2012). These findings on NRT2.1 closely resemble what is currently known for *H. polymorpha* NRT2-family protein (YNT1) regulation. Phorphorylation in the long cytosolic loop of this fungal transporter is required for its membrane sorting. This phosphorylation happens under carbon sufficient conditions in presence of nitrate, but also in absence of it. Interestingly, this membrane localization of YNT1 is required precisely for rapid induction of other nitrate utilization genes (Navarro et al., 2008; Martin et al., 2012). When joining the recent information from the two NRT2 class transporters, the NRT2.1 of *Arabidopsis* and YNT1 from *H. polymorpha*, the picture of NTR2-type nitrate-uptake independent signaling role of these transporters starts to merge. In both of these eukaryotes correct membrane trafficking of the transporter protein under carbon sufficient conditions and in absence of repressive nitrogen source is a fundamental pre-requirement for their intracellular nitrate uptake-independent signal transducer activities. The control of this membrane localization is carried out either via direct phosphorylation of the transporter and/or posttranslational modifications of its interactive proteins. Our results suggest that a certain cellular threshold level of the *Laccaria* transporter is equally required for triggering the rest of the nitrate utilization pathway. Moreover, this regulatory role appears to be independent of the presence of exogenous nitrate. Taking in account the novel data on nitrate uptake-independent signaling function of NRT2-type transporters, the co-downregulation effects of *LbNrt*-silencing could be explained by action of a highly similar cellular mechanism. As not responsive to nitrate induction, capable of simultaneous uptake of nitrate and organic nitrogen, and apparently only controlled by the presence/absence of repressive nitrogen source, ammonium, a predominant membrane localization of the LbNRT would be expected under all carbon sufficient and fHANT-AC non-repressive growth conditions. Also multiple posttranslational modification motifs in the protein suggest that, besides regulated at transcriptional level, the membrane trafficking of the LbNRT should play a significant role in fHANT-AC fine-tune control. Therefore, we postulate that a simple insufficient membrane accumulation of the transporter due to efficient RNAi silencing could be behind initiating an intracellular LbNRT-linked but nitrate-independent signaling cascade. This would lead to significant reduction of *LbNr* and to lesser extent of *LbNir* transcripts in the cell either via modifying the transcriptional activity and/or affecting the mRNA stability of these fHANT-AC genes. Such transporter level-dependent regulation may result important for flexible activation of the nitrate utilization pathway and consequently results in maximal cellular metabolization of nitrate under temporally and spatially highly variable nitrogen growth condition as is forest soil.

Since *Laccaria bicolor* counts with only one nitrate transporter protein (*versus* multiple transporters in plants and two paralogues in *H. polymorpha* and *A. nidulans*) it offers a unique system for further studies on the regulatory role of this protein and nitrate metabolism in ECM fungi and in eukaryotes, in general. Our results also strongly support the use of RNAi in manipulation of other fungal nutrient transporters, specially the ones predicted to be fundamental in nutrient transfer between the fungus and the host plant.

Acknowledgments

This work was supported by Universidad Nacional de Quilmes, CONICET and Agencia Nacional de Promoción Científica y Tecnológica (ANPCyT), Argentina.

References

Alder, M. N., Dames, S., Gaudet, J., and Mango, S. E. (2003) Gene silencing in *Caenorhabditis elegans* by transitive RNA interference. *RNA* 9: 25-32.

Andrianopoulos, A., Kourambas, S., Sharp, J. A., Davis, M. A., and Hynes, M. J. (1998) Characterization of the *Aspergillus nidulans nmrA* gene involved in nitrogen metabolite repression. *J. Bacteriol.* 180: 1973-1977.

Aouadj, R., Es-Sgaouri, A. and Botton, B. (2000) A study of the stability and properties of nitrate reductase from the ectomycorrhizal fungus *Pisolithus tinctorius*. *Cryptogamie Mycologie* 21: 187-202.

Avolio, M., Müller, T., Tpangara, A., Fitz, M., Becker, B., Pauck, A., Kirsch, A., and Wipf, D. (2012) Regulation of genes involved in nitrogen utilization on different C/N ratios and nitrogen sources in the model ectomycorrhizal fungus *Hebeloma cylindrosporum*. *Mycorrhiza* Epub. ahead of print.

Bailly, J., Debaud, J.-C., Vegner, M.-C., Plassard, C., Chalot, M., Marmeisse, R., and Fraissinet-Tachet, L., (2007) How does a symbiotic fungus modulate expression of its host-plant nitrite reductase? *New Phytol.* 175: 155-165.

Berger, H., Basheer, A., Böck, S., Reyes-Dominguez, Y., Dalik, T., Altmann, F., and Strauss, J. (2008) Dissecting individual steps of nitrogen transcription factor cooperation in the *Aspergillus nidulans* nitrate cluster. *Mol. Microbiol.* 69: 1385-1398.

Berger, H., Pachlinger, R., Morozov, I., Goller, S., Narendja, F., Caddick, M., and Strauss, J. (2006) The GATA factor AreA regulates localization and in vivo binding site occupancy of the nitrate activator NirA. *Mol. Microbiol.* 59: 433-446.

Bernreiter, A., Ramon, A., Fernandez-Martinez, J., Berger, H., Araujo-Bazon, L., Esperso, E. A., Pachlinger, R., Gallmetzer, A., Anderl, I., Scazzocchio, C., and Strauss, J. (2007) Nuclear export of the transcription factor NirA is a regulatory checkpoint for nitrate induction in *Aspergillus nidulans*. *Mol. Cell. Biol.* 27: 791-802.

Blew, R. G. and Parkinson, D. (1993) Nitrification and denitrification in a while spruce forest in southwest Alberta, Canada. *Can. J. Forest Res*. 23: 1715-1719.

Booth, M. S., Stark, J. M. and Rastetter, E. (2005). Controls on nitrogen cycling in terrestrial ecosystems: a synthetic analysis of literature data. *Ecol. Monographs* 75: 139-157.

Butler, S. M., Malillo, J. M., Johnson, J. E., Mohan, J., Steudler, P. A., Lux, H., Burrows, E., Smith, R. M., Vario, C. L., Scott, L., Hill, T. D., Aponte, N., and Bowles, F. (2012) Soil warming alters nitrogen cyclin in a New England forest: implications for ecosystem function and structure. *Oecologia* 168: 819-828.

Cabello, P., Roldán, M. D. and Moreno-Vivián, C. (2004) Nitrate reduction and the nitrogen cycle in archaea. *Microbiol.* 150: 3527-3546.

Caddick, M. X., Jones, M. G., van Tonder, J. M., Le Cordier, H., Narendja, F., Strauss, J., and Morozov, I. Y. (2006) Opposing signals differentially regulate transcript stability in *Aspergillus nidulans*. *Mol. Microbiol.* 62: 509-519.

Caddick, M. X., Peters, P. and Platt, A. (1994) Nitrogen regulation in fungi. *Antonie* van *Leewenhoek* 65: 169-177.

Chen, X., Stone, M., Schlagnhaufer, C., Romaine, C. P. (2000) A fruiting body tissue method for efficient *Agrobacterium*-mediated transformation of *Agaricus bisporus*. *Appl. Environ. Microbiol.* 66: 4510-4513.

Crawford, N. M. and Arst, H. N. Jr. (1993) The molecular genetics of nitrate assimilation in fungi and plants. *Ann. Rev. Genet.* 27: 115-146.

Dechorgnat, J., Nguyen, C. T., Armengaud, P., Jossier, M., Diatloff, Filleur, S., and Daniel-Vedele, F. (2011) From the soil to the seeds: the long journey of nitrate in plants. *J. Exp. Bot.* 62: 1249-1359.

Fernandez, E. and Galvan, E. (2008) Nitrate assimilation in *Chlamydomonas*. *Euk. Cell* 7: 555-559.

Forde, B. G. (2000) Nitrate transporters in plants: structure, function and regulation. *Biochim. Biophys. Acta* 1465: 219-235.

Fu, Y. H. and Marzluf, G. A. (1987) Characterization of *nit-2*, the major nitrogen regulatory gene of *Neurospora crassa*. *Mol. Cell. Biol.* 7: 1691-1696.

Fu, Y.-H., Feng, B., Evans, S., and Marzluf, G. A. (1995) Sequence-specific DNA binding by NIT4, the pathway specific regulatory protein that mediates nitrate induction in *Neurospora*. *Mol. Microbiol.* 15: 935-942.

Galván, A. and Fernández, E. (2001) Eukaryotic nitrate and nitrite transporters. *Cell. Mol. Life Sci.* 58: 225-233.

Gao-Rubinelli, F. and Marzluf, G. A. (2004) Identification and characterization of a nitrate transporter gene in *Neurospora crassa*. *Biochem. Genet.* 42: 21-34.

Gobert, A. and Plassard, C. (2002) Differential NO_3^- dependent patterns of NO_3^- uptake in *Pinus pinaster*, *Rhizopogon roseolus* and their ectomycorrhizal association. *New Phytol.* 154: 509-516.

Gobert, A. and Plassard, C. (2007) Kinetics of NO_3^- net fluxes in *Pinus pinaster*, *Rhizopogon roseolus* and their ectomycorrhizal association, as affected by the presence of NO_3^- and NH_4^+. *Plant Cell Environ.* 30: 1309-1319.

Gojon, A., Krouk, G., Perrine-Walker, F., and Laugier, E. (2011) Nitrate transceptor(s) in plants. *J. Exp. Bot.* 62: 2299-2308.

Gorfer, M., Blumhoff, M., Klaubauf, S., Urban, A., Inselsbacher, E., Bandian, D., Mitter, B., Sessitsch, A., Wanek, W., and Strauss, J. (2011) Community profiling and gene expression of fungal assimilatory nitrate reductases in agricultural soil. *ISME J.* 11: 1771-1783.

Guescini, M., Pierloeni, R., Palma, F., Zeppa, S., Vallorani, L., Potenza, L., Sacconi, G., Giomaro, G., and Stocchi, V. (2003). Characterization of the

Tuber borchii nitrate reductase gene and its role in ectomycorrhizae. *Mol. Genet. and Genom.* 269: 807-816.

Guescini, M., Zeppa, S., Pierleoni, R., Sisti, D., Stocchi, L., and Stocchi, V. (2007) The expression profile of the *Tuber borchii* nitrite reductase suggests its positive contribution to the host plant nitrogen nutrition. *Curr. Genet.* 5: 31–41.

Haley, G., Foys, B. and Levine, M. (2010) Vectors and parameters that enhance the efficacy of RNAi-mediated gene distruption in transgenic *Drosophila*. *Proc. Nat. Acad. Sci. US* 107: 11435-11440.

Ho, C. H., Lin, S. H., Hu, H. C., and Tsay, Y. F. (2009) CHL1 functions as a nitrate sensor in plants. *Cell* 138: 1184-94.

Hofmann, K. and Stoffel, W. (1993) TMbase - A database of membrane spanning proteins segments. *Biol. Chem. Hoppe-Seyler* 374: 166.

Houmard, N. M., Mainville, J. L., Bonin, C. P., Huang, S., Luethy, M. H., and Malvar, T. M. (2007) High-lysine corn generated by endosperm-specific suppression of lysine cabolism using RNAi. *Plant Biotechnol. J.* 5: 605-615.

Jackson, A. L., Burchard, J., Schelter, J., Chau, B. N., Cleary, M., Lim, L., and Linsley, P. S. (2006) Widespread siRNA "off-target" transcript silencing mediated by seed region sequence complementarity. *RNA* 12: 1179-1187.

Jargeat, P., Gay, G., Debaud, J. C., and Marmeisse, R. (2000) Transcription of a nitrate reductase gene isolated from the symbiotic basidiomycete fungus *Hebeloma cylindrosporum* does not require induction by nitrate. *Mol. Gen. Genom.* 263: 948–956.

Jargeat, P., Rekangalt, D., Verner, M., Gay, G., DeBaud, J., Marmeisse, R., and Fraissinet-Tachet, L. (2003) Characterisation and expression analysis of a nitrate transporter and nitrite reductase genes, two members of a gene cluster for nitrate assimilation from the symbiotic basidiomycete *Hebeloma cylindrosporum*. *Curr. Genet.* 43: 199–205.

Jerabkova, L., Prescott, C. E. and Kishchuk, B. E. (2006) Nitrogen availability in soil and forest floor of contrasting types of boreal mixedwood forests. *Can. J. Forest Res.* 36: 112-122.

Kemppainen, M. J., Alvarez Crespo, M. C. and Pardo, A. G. (2010) fHANT-AC genes of the ectomycorrhizal fungus *Laccaria bicolor* are not repressed by L-glutamine allowing simultaneous utilization of nitrate and organic nitrogen sources. *Environ. Microbiol. Rep.* 2: 541-553.

Kemppainen, M., Circosta, A., Tagu, D., Martin, F., and Pardo, A. G. (2005) *Agrobacterium*-mediated transformation of the ectomycorrhizal symbiont *Laccaria bicolor* S238N. *Mycorrhiza* 16: 19–22.

Kemppainen, M., Duplessis, S., Martin, F., and Pardo, A. G. (2008) T-DNA insertion, plasmid rescue and integration analysis in the model mycorrhizal fungus *Laccaria bicolor*. *Microb. Biotechnol.* 1: 258–269.

Kemppainen, M., Duplessin, S., Martin, F., and Pardo, A. G. (2009) RNA silencing in the model mycorrhizal fungus *Laccaria bicolor*: gene knock-down of nitrate reductase results in inhibition of symbiosis with *Populus*. *Environ. Microbiol.* 11: 1878–1896.

Kemppainen, M. and Pardo, A. G. (2010) pHg/pSILBAγ vector system for efficient gene silencing in homobasidiomycetes: optimization of ihpRNA-triggering in the mycorrhizal fungus *Laccaria bicolor*. *Microb. Biotechnol.* 3: 178–200.

Kemppainen, M. J. and Pardo, A. G. (2013) *LbNrt* RNA silencing in the mycorrhizal symbiont *Laccaria bicolor* reveals a nitrate-independent regulatory role for a eukaryotic NRT2-type nitrate transporter. *Environ. Microbiol. Rep.* 5: 353-366.

Krogh, A., Larsson, B., con Heijne, G., and Sonnhammer, E. L. (2001) Predicting transmembrane protein topology with a hidden Markov model: application to complete genome. *J. Mol. Biol.* 305: 567-580.

Kudla, B., Caddick, M. X., Langdon, T., Marinez-Rossi, N. M., Bennett, C. F., Sibley, C. F., Davies, R. W., and Arst, H. N. (1990) The regulatory gene *areA* mediating nitrogen metabolite repression in *Aspergillus nidulans*. Mutations affecting specificity of gene activation alter a loop residue of a putative zinc finger. *EMBO J.* 9: 1355-1364.

Langdon, T., Sheerins, A., Ravagnani, A., Gielkens, M., Caddick, M. X., and Arst, H. N. Jr. (1995) Mutational analysis reveals dispensability of the N-terminal region of the *Aspergillus* transcription factor mediating nitrogen metabolite repression. *Mol. Microbiol.* 17: 877-888.

Laugier, E., Bouguyon, E., Mauriés, A., Tillard, P., Gojon, A., and Lejay, L., 2012. Regulation of high-affinity nitrate uptake in roots of *Arabidopsis* depends predominantly on posttranscriptional control of the NRT2.1/ NAR2.1 transport system. *Plant Physiol.* 158: 1067-1078.

Leverman, A. M., Zoomer, H. R., van Versevald, H. W., and Verhoef, H. A. (2000) Temporal and spatial variation of nitrogen transformations in a coniferous forest soils. *Soil Biol. Biochem.* 32: 1661–1670.

Little, D. Y., Rao, H., Oliva, S., Daniel-Vedele, F., Krapp, A., and Malamy, J. E. (2005) The putative hight-affinity nitrate transporter NRT2.1 repressed lateral root initiation in response to nutritional cues. *Proc. Nat. Acad. Sci. US* 102: 13693-13698.

Liu, K. H., Tsay, Y. F. (2003) Switching between the two action modes of the dual-affinity nitrate transporter CHL1 by phosphorylation. *EMBO J.* 22: 1005-1013.

Lucic, E., Fourrey, C., Kohler, A., Martin, F., Chalot, M., and Brun-Jacob, A. (2008) A gene repertoire for nitrogen transport in *Laccaria bicolor*. *New Phytol.* 180: 323–364.

Luque-Almagro, V. M., Gates, A. J., Moreno-Vivián, C., Ferguson, S. J., Richardson, D. J., and Roldán, M. D. (2011) Bacterial nitrate assimilation: gene distribution and regulation. *Biochem. Soc. Trans.* 39: 1838-1843.

Machín, F., Perdomo, G., Pérez, M. D., Brito, N., and Siverio, J. M. (2000) Evidence for multiple nitrate uptake systems in *Hansenula polymorpha*. *FEMS Microbiol. Lett.* 194: 171-174.

Malamy, J. E. and Ryan, K. S. (2001) Environmental regulation of lateral root initiation in *Arabidopsis*. *Plant Physiol.* 127: 899-909.

Martin, F., Aerts, A., Ahrén, D., Brun, A., Danchin, E. G., Duchaussoy, F., et al. (2008) The genome of *Laccaria bicolor* provides insights into mycorrhizal symbiosis. *Nature* 452: 88–92.

Martin, F., Kohler, A., Murat, C., Balestrini, R., Couinho, P. M., Jaillon, O., et al. (2010) Périgord black truffle genome uncovers evolutionary origins and mechanisms of symbiosis. *Nature* 464: 1033-1038.

Martín, Y., González, Y. V., Cabrera, E., Rodríguez, C., and Siverio, J. M. (2011) Npr1 Ser/Thr protein kinase links nitrogen source quality and carbon availability with the yeast nitrate transporter (Ynt1) levels. *J. Biol. Chem.* 286: 27225-27235.

Marzluf, G. A. (1997) Genetic regulation of nitrogen metabolism in the fungi. *Microbiol. Mol. Biol. Rev.* 61: 17–32.

McClure, P. R., Kochian, L. V., Spanswick, R. M., and Shaff, J. E. (1990a) Evidence for cotransport and protons in maize roots: I. Effects of nitrate on the membrane potential. *Plant Physiol.* 93: 281-289.

McClure, P. R., Kochian, L. V., Spanswick, R. M., and Shaff, J. E. (1990b) Evidence for cotransport of nitrate and protons in maize roots: II. Measurement of NO(3) and H fluxes with the ion-selective microelectrodes. *Plant Physiol.* 93: 290-294.

Miller, A. J., Fan, X., Orsel, M., Smith, S. J., and Wells, D. M. (2007) Nitrate transport and signalling. *J. Exp. Bot.* 58: 2297–2306.

Mitter, N., Mitchell, R. and Dietzgen, R. G. (2006) Fate of hairpin transcript components during RNA silencing and its suppression in trangenic viruresitant tobacco. *J. Biotechnol.* 126: 115-122.

Mo, X. K. and Marzluf, G. A. (2003) Cooperative action of the NIT2 and NIT4 transcription factors upon gene expression in *Neurospora crassa. Curr. Genet.* 43: 260-267.

Montanini, B., Viscomi, A. R., Bolchi, A., Martin, Y., Siverio, J. M., Balestrini, R., Bonfante, P., and Ottonello, S. (2006a) Functional properties and differential mode of regulation of the nitrate transporter from a plant symbiotic ascomycete. *Biochem. J.* 394: 125–34.

Montanini, B., Gabella, S., Abbà, S., Peter, M., Kohler, A., Bonfante, P., Chalot, M., Martin, F., and Ottonello, S. (2006b) Gene expression profiling of the nitrogen starvation stress response in the mycorrhizal ascomycete *Tuber borchii. Fungal Genet. Biol.* 43: 630–641.

Morozov, I. Y., Galbis-Martinez, M., Jones, M. G., and Caddick, M. X. (2001) Characterization of nitrogen metabolite signaling in *Aspergillus* via the regulated degradation of area mRNA. *Mol. Microbiol.* 42: 269-277.

Mullins, E. D., Chen, X., Romaine, C. P., Raina, R., Geiser, D. M., and Kang, S. (2001) *Agrobacterium*-mediated transformation of *Fusarium oxysporum*: an efficient tool for insertional mutagenesis and gene transfer. *Phytopathol.* 91: 173-180.

Muro-Pastor, M. I., Strauss, J., Ramon, A., and Scazzocchio, C. (2004) A paradoxical mutant GATA factor. *Euk. Cell* 3: 394-405.

Myrold, D. D. and Ritchie Posavatz, N. (2008) Potential importance of bacteria and fungi in nitrate assimilation in soil. *Soil Biol. Biochem.* 39: 1737-1743.

Narendja, F., Goller, S. P., Wolschek, M., and Strauss, J. (2002) Nitrate and the GATA factor AreA are necessary for in vivo binding of NirA, the pathway-specific transcriptional activator of *Aspergillus nidulans*. *Mol. Microbiol.* 44: 573-583.

Navarro, F. J., Machín, F., Martín, Y., and Siverio, J. M. (2006) Down-regulation of eukaryotic nitrate transporter by nitrogen-dependent ubiquitinylation. *J. Biol. Chem.* 281: 13268-13274.

Navarro, F. J., Martín, Y. and Siverio, J. M. (2008) Phosphorylation of the yeast nitrate transporter Ynt1 is essential for delivery to the plasma membrane during nitrogen limitation. *J. Biol. Chem.* 283: 31208-31217.

Navarro, F. J., Perdomo, G., Tejera, P., Medina, B., Machín, F., Guillén, R. M., Lancha, A., and Siverio, J. M. (2003) The role of nitrate reductase in the regulation of the nitrate assimilation pathway in the yeast *Hansenula polymorpha. FEMS Yeast Res.* 4: 149-155.

Nguyen, Q. B., Kadotani, N., Kasahara, S., Tosa, Y., Mayama, S., and Nakayashiki, H. (2008) Systematic functional analysis of calcium-signalling proteins in the genome of the rice-blast fungus, *Magnaporthe oryzae*, using a high-throuhgput RNA-silencing system. *Mol. Microbiol.* 68: 1348-1365.

Nishimura, K., Hirai, S. and Kodoma, H. (2009) Effects of siRNAs targeting the spacer sequence of plant RNAi vectors on the specificity and efficiency of RNAi. *J. Biosc. Bioeng.* 108: 435-437.

Nygren, C. M. R., Eberhardt, U., Karlsson, M., Parret, J. L., Lindahl, B. D., and Taylor, A. F. S. (2008) Growth on nitrate and occurrence of nitrate reductase-encoding genes in a phylogenetically diverse range of ectomycorrhizal fungi. *New Phytol.* 180: 875–889.

Okamoto, M., Kumar, A., Li, W., Wang, Y., Siddiqi, M. Y., Crawford, N. M., and Glass, A. D. M. (2006) High-affinity nitrate transport in root of *Arabidopsis* depends on expression of the NAR2-like gene ArNRT3.1. *Plant Physiol.* 140: 1036-1046.

Orsel, M., Chopin, F., Leleu, O., Smith, S. J., Krapp, A., Daniel-Vedele, F., and Miller, A. J. (2006) Characterization of a two component high affinity nitrate uptake system in *Arabidopsis*; physiology and protein-protein interaction. *Plant Physiol.* 142: 1304-1317.

Pao, S. S., Paulsen, I. T. and Saier, M. H. Jr. (1998) Major facilitator superfamily. *Microbiol. Mol. Biol. Rev.* 62: 1-34.

Plett, J. M. and Martin, F. (2011) Blurred boundaries: lifestyle lessons from ectomycorrhizal fungal genomes. *Trends Genet.* 27: 14-22.

Plett, J. M., Kemppainen, M., Kale, S. D., Kohler, A., Legué, V., Brun, A., Tyler, B. M., Pardo, A. G., and Martin, F. (2011) A secreted effector protein of *Laccaria bicolor* is required for symbiosis development. *Curr. Biol.* 21: 1197-1203.

Punt, P. J., Strauss, J., Smit, R., Kinghorn, J. R., van den Honderl, C. A. M. J. J., and Scazzocchio, C. (1995) The intergenic region between the divergently transcribed *niiA* and *niaD* genes of *Aspergillus nidulans* contains multiple NirA-binding sites which act bidirectionally. *Mol. Cell. Biol.* 10: 5688-5699.

Rékangalt, D., Pépin, R., Verner, M.-C., Debaud, J.-C., Marmeisse, R., Fraissinet-Tachet, L. (2009) Expression of the nitrate transporter *nrt2* gene from the symbiotic basidiomycete *Hebeloma cylindrosporum* is affected by host plant and carbon sources. *Mycorrhiza* 19: 143-148.

Remans, T., Nacry, P., Pervent, M., Girin, T., Tillard, P., Lepetit, M., and Gojon, A. (2006) A central role for the nitrate transporter NRT2.1 in the integrated morphological and physiological responses of the root system to nitrogen limitation in *Arabidopsis*. *Plant Physiol.* 140: 909-921.

Rennenberg, H., Dannenmann, M., Gessler, A., Kreuzwieser, J., Simon, J., and Papen, H. (2009) Nitrogen blance in forest soils: nutritional limitation of plants under climate change stresses. *Plant Biol.* 11: 4-23.

Rexach, J., Llamas, A., Fernández, E., Galván, A. (2002) The activity of the high-affinity nitrate transport system I (NRT2;1, NAR2) is responsible for the efficient signaling of nitrate assimilation genes in *Clamydomonas reinhardtii*. *Planta* 215: 606-611.

Rossi, B. and Berardi, E. (2009) Assimilatory nitrate reduction in *Hansenula polymorpha*. *Yeast Biotechnology: Diversity and Applications I*, 307-320. DOI: 10.1007/978-1-4020-8292-4_15.

Sarbhoy, A. K. (1965) Nutritional studies on some members of the mucorales. V. Utilization of different nitrogen compounds. *Pathologia et Microbiologia* 28: 816–829.

Scazzocchio, C. (2000) The fungal GATA-factors. *Curr. Op. Microbiol.* 3: 126-131.

Scheromm, P., Plassard, P. and Salsac, L. (1990) Regulation of nitrate reductase in the ectomycorrhizal basidiomycete *Hebeloma cylindrosporum* Romagn., cultured on nitrate or ammonium. *New Phytol.* 112: 441-447.

Schinko, T., Berger, H., Lee, W., Gallmetser, A., Pirker, K., Pachllnger, R., Buchner, I., Relchenauer, T., Gükdener, U., and Strauss, J. (2010) Transcriptome analysis of nitrate assimilation in *Aspergillus nidulans* reveals connection to nitric oxide metabolism. *Mol. Microbiol.* 78: 720-738.

Schleppi, P., Bucher-Wallin, I., Hagedorn, F., and Körner, C. (2012) Increased nitrate availability in the soil of a mixed mature temperate forest subjected to elevated CO_2 concentration (canopy FACE). *Global Change Biol.* 18: 757-758.

Siverio, J. M. (2002) Assimilation of nitrate by yeasts. *FEMS Microbiol. Rev.* 26: 277-284.

Slot, J. C. and Hibbett, D. S. (2007) Horizontal transfer of a nitrate assimilation gene cluster and ecological transition in fungi: a phylogenetic study. *PLoS One* 2: e1097.

Slot, J. C., Hallstrom, K. N., Matheny, P. B., and Hibbett, D. S. (2007) Diversification of NRT2 and the origin of its fungal homolog. *Mol. Biol. Evol.* 24: 1731–1743.

Slot, J. C., Hallstrom, K. N., Matheny, P. B., Hosaka, K., Mueller, G., Robertson, D. L., and Hibbett, D. S. (2010). Phylogenetic, structural and functional diversification of nitrate transporters in the three ecologically diverse clades of mushroom-forming fungi. *Fungal Ecol.* 3: 160-177.

Smith, S. and Read, D. J. (2008) *Mycorrhizal Simbiosis*. London, UK: Academic Press.

Stark, J. M. and Har, S. C. (1997) High rates of nitrification and nitrate turn-over in undisturbed coniferous forests. *Nature* 385: 61-64.

Strahm, B. D. and Harrison, R. B. (2006) Nitrate sorption in a variable-charge forest soil of the Pacific Northwest. *Soil Sci.* 171: 313-321.

Tamm, C. O. (1991). *Nitrogen in terrestrial ecosystems. Questions of productivity, vegetation changes, and ecosystem stability*. Berlin, Germany: Springer-Verlag.

Tamura, K., Dudley, J., Nei, M., and Kumar, S. (2007) MEGA4: Molecular Evolutionary Genetics Analysis (MEGA) software version 4.0. *Mol. Biol. Evol.* 24: 1596-1599.

Tao, Y. and Marzluf, G. A. (1999). The NIT2 nitrogen regulatory protein of *Neurospora*: expression and stability of *nit-2* mRNA and protein. *Curr. Genet.* 36: 153-158.

Todd, R. B., Fraser, J. A., Wong, K. H., Davis, M. A., and Hynes, M. J. (2005) Nuclear accumulation of the GATA factor AreA in response to complete nitrogen starvation by regulation of nuclear export. *Euk. Cell* 4: 1646–1653.

Tomsett, A. B., Dunn-Coleman, N. S. and Garrett, R. H. (1981) The regulation of nitrate assimilation in *Neurospora crassa*: the isolation and genetic analysis of *nmr-1* mutants. *Mol. Gen. Genet.* 182: 229-233.

Trueman, L. J., Richardson, A. and Forbe, B. G. (1996) Molecular cloning of higher plant homologues of the high-affinity nitrate transporters of *Chlamydomonas reinhardtii* and *Aspergillus nidulans*. *Gene* 175: 223-231.

Unkles, S. E., Rouch, D. A., Wang, Y., Siddiqi, M. Y., Glass, A. D., and Kinghorn, J. R. (2004) Two perfectly conserved arginine residues are required for substrate binding in a high-affinity nitrate transporter. *Proc. Nat. Acad. Sci. US* 101: 17549–17554.

Unkles, S. E., Zhou, D., Siddiqi, M. Y., Kinghorn, J. R., and Glass, A. D. (2001) Apparent genetic redundancy facilitates ecological plasticity for nitrate transport. *EMBO J.* 20: 6246-6255.

Von Heijne, G. (1992) Membrane protein structure prediction. Hydrophobicity analysis and the positive-inside rule. *J. Mol. Biol*. 225: 487-494.

Wang, Y., Li, W., Siddiqi, Y., Kinghorn, J. R., Unkles, S. E., and Glass, A. D. M. (2007) Evidence for post-translational regulation of NrtA, the *Aspergillus nidulans* high-affinity nitrate transporter. *New Phytol.* 175: 699–706.

Wong, K. H., Hynes, M. J. and Davis, M. A. (2008) Recent advances in nitrogen regulation: a comparison between *Saccharomyces cerevisiae* and filamentous fungi. *Euk. Cell* 7: 917–925.

Wong, K. H., Hynes, M. J., Todd, R. B., and Davis, M. A. (2007). Transcriptional control of nmrA by the bZIP transcription factor MeaB reveals a new level of nitrogen regulation in *Aspergillus nidulans*. *Mol. Microbiol.* 66: 534-551.

Wong, K. H., Hynes, M. J., Todd, R. B., and Davis, M. A. (2009) Deletion and overexpression of the Aspergillus nidulans GATA factor AreB reveals unexpected pleiotropy. *Microbiol.* 155: 3868-3880.

Yamanaka, T. (1999) Utilization of inorganic nitrogen in pure cultures by saprotrophic and ectomycorrhizal fungi producing sporophores on urea-treated forest floor. *Mycol. Res.* 103: 811–816.

Zhou, J. J., Trueman, L. J., Boorer, K. J., Theodoulou, F. L., Forde, B. G., and Mille, A. J. (2000) A high affinity fungal nitrate carrier with two transport mechanisms. *J. Biol. Chem.* 275: 39894-39899.

In: Mycorrhizas
Editor: Elaine Warwick
ISBN: 978-1-63321-436-1

Chapter 5

ARBUSCULAR MYCORRHIZAL FUNGI IN THE ALLEVIATION OF OXIDATIVE STRESS UNDER CUTTING PROPAGATION MANAGEMENT

María Josefina Bompadre [1], Roxana Colombo[2], María del Carmen Ríos de Molina[3], Alicia Margarita Godeas[2] and Alejandro Guillermo Pardo[1,*]

[1] Laboratorio de Micología Molecular, Departamento de Ciencia y Tecnología, Universidad Nacional de Quilmes (UNQ), and Consejo Nacional de Investigaciones Científicas y Técnicas (CONICET), (B1876BXD), Bernal, Provincia de Buenos Aires, Argentina

[2] Departamento de Biodiversidad y Biología Experimental. Facultad de Ciencias Exactas y Naturales, Universidad de Buenos Aires (UBA). Intendente Güiraldes, Buenos Aires, Argentina

[3] Instituto de Química Biológica de la Facultad de Ciencias Exactas y Naturales (IQUIBICEN), Universidad de Buenos Aires (UBA), 4to piso, Pabellón 2. C1428EGA. Buenos Aires, Argentina

* Corresponding author: Alejandro G. Pardo, Phone: 54-11-4365-7100 (ext 5639), Fax: 54-11-4365-7132, E-mail: apardo@unq.edu.ar.

ABSTRACT

The capacity of roots to sense soil physicochemical parameters plays an essential role in maintaining plant nutritional and developmental functions under abiotic stress. Drought greatly affects the growth and development of plants and can cause extensive losses to agricultural productivity. The water regime affects a wide variety of physiological and biochemical processes in plants including an increased production of reactive oxygen species (ROS) capable of causing oxidative damage to proteins, DNA and lipids. Moreover, this ROS production can induce cellular, anatomical, and morphological changes that improve drought tolerance. Arbuscular mycorrhizal fungi (AMF) colonize a wide range of plant species though the ability of different AMF to promote host growth and/or water deficit resistance is variable. AMF contributes to drought tolerance through a combination of physical, nutritional and physiological effects. The response to oxidative stress depends on many factors, including the organism tissue and the degree of stress. In most cases mycorrhizal plants increase antioxidant defenses, such as the ROS-scavenging enzymes superoxide dismutase (SOD), catalase (CAT), and ascorbate peroxidase (APX). AMF can maintain the levels of malondialdehyde (MDA), a compound indicative of lipid damage, by rapidly increasing antioxidant defenses and preventing lipid damage. In other cases an increase of MDA is observed in mycorrhizal plants due to an active response to oxidative stress. A critical step of AMF inoculation technology is the appropriate selection of fungal isolates to be used as inoculants. Nursery cultivation is the most convenient way to improve the success of semi-woody cuttings transplantation (e.g. olive, pomegranate, citrus) at the crop area. Plants suffer at least two transplant moments in nursery conditions prior to outside cultivation. AMF can reduce transplant stress by changing the morphology of the root system favouring the establishment of plants. Thus AMF can greatly contribute to crop productivity and environmental sustainability. Artificial inoculation of plant cuttings with AMF at an early stage of cutting development has been adopted by an increasing number of nursery managers as a method for promoting growth, production and precocity. Moreover, it is an essential component for most plants and it can be used as a biofertilizer resource. In this chapter we discuss the effect of an early AMF inoculation in alleviating oxidative stress in semi-woody cutting propagation management under nursery conditions. We also discuss the benefits of AMF to solve the problematic issue of water deficit and transplant stresses generated during plant growth.

WATER DEFICIT AND PLANT GROWTH

Life on Earth has always depended on the water. It is the natural habitat of a large part of the organisms and plays an important role in the growth and development of plants. Water is a solvent for many organic and inorganic substances and is the medium in which the enzymatic reactions take place. Is the major component of plants as it is responsible for 80-90% of the fresh weight in herbaceous plants and more than 50% of woody plants (Nilsen and Orcutt, 1996).

Water deficit is the most important factor that causes a reduction in plant growth and they affects a variety of physiological and biochemical processes in plants (Sanchez-Diaz and Aguirreolea, 2002). Thus the loss of water in a plant tissue has the following effects which can influence the metabolism of the plant:

- Reduced water potential.
- Decreased turgor.
- Concentration of molecules, as the cell volume decreases with reduced turgidity.
- Impaired spatial relationships in the plasmalemma, tonoplast membranes and organelles due to volume changes.
- Changing the structure or configuration of the macromolecules due to the removal of water of hydration.

The most important effect of mild water stress is the reduction in growth, being particularly sensitive cell expansion and a lesser extent cell division. In addition to growth inhibition, water deficits alter development and plant morphology by changing the stem / root ratio. Turn it affects the productivity of plants (Wang et al., 2001; Franco et al., 2011). Other effects on plant development is premature termination of growth in length of perennials with a resting bud formation. Water deficits also cause abscission of leaves and fruits, especially after the disappearance of stress. Also affect reproductive development, necessitating, in some species, a drought to stimulate floral initiation or cause the emergence of flower buds and differentiated. Depending on the amount of water loss, the loss rate and the duration of the stress conditions, may result in solute concentration, change in the volume and shape of the membranes, disruption of water potential gradient, loss of turgor, disruption of membrane integrity and denaturation of protein structure or function (Bray, 1997; Smirnoff, 1998).

OXIDATIVE STRESS, INDUCTION OF ANTIOXIDANT PROTEINS

Not only the lack of water but also other adverse environmental conditions exert its negative effect, in turn generating oxidative stress. The responses vary in their speed (Bray, 1997), they may occur within a few seconds (e.g. through the change in the phosphorylation state of a protein) or within a few minutes to hours (through changes in gene expression).

The stress response is initiated when the plant recognizes the stress at the cellular level, then signal transduction pathways which transmit information between cells and through the plant are activated. Some genes were strongly expressed while others are repressed (Bray et al., 2000). The functions of these proteins and the mechanisms regulating its expression are currently as central studies (Porcel et al., 2003; Roldán et al., 2008; Miller et al., 2010; Ruíz-Lozano et al., 2012).

In plants, reactive oxygen species (ROS) metabolism: superoxide anion radical ($O_2^{\cdot -}$), hydrogen peroxide (H_2O_2) and hydroxyl radical ($\cdot OH$) are in a dynamic balance. Under water stress conditions the balance is broken and the antioxidant system need to reduce the damage to the tissues (Wu et al., 2006b). This imbalance may be caused by overproduction of ROS or by a deficiency in antioxidant system (Morel and Barouki, 1999).

To overcome the deleterious effect of ROS, there are enzymatic and non-enzymatic antioxidants that help plants to enhance stress tolerance. Both hydrogen peroxide (H_2O_2) and the superoxide anion radical ($O_2^{\cdot -}$) are formed in the electron transport chain of mitochondria and chloroplasts, and by many dependent flavin oxidases, including glycolate oxidase peroxisome (Del Rio et al., 1992; Mittler, 2002). H_2O_2 and $O_2^{\cdot -}$ they are also generated by non-enzymatic mechanisms as iron catalyzed oxidation of NAD(P)H, reduced glutathione (GSH) and ascorbate. The H_2O_2 can inactivate several enzymes of the Calvin cycle (Asada and Takahashi, 1987; Noctor and Foyer, 1998). It is estimated that 10-20% of the electrons that pass through photosystem I reduce O_2 to H_2O_2 (Asada, 1999) and that 2% of the O_2 consumed by mitochondria is used to generate H_2O_2 (Scandalios et al., 1997) therefore, plants and other aerobic organisms depend on antioxidant defense to prevent their own destruction (Rizhsky et al. 2003).

Plants have shared a problem with all other aerobic organisms. Molecular oxygen (O_2) is required for energy metabolism demand, but the presence of O_2 has risks of oxidative damage due to the formation of active free radicals such as $O_2^{\cdot-}$, H_2O_2 and ·OH (Dalton, 1995). Photosynthesis can be seen as a risk by the amount of O_2 that occur in the vicinity of a redox system, which can reduce O_2 a $O_2^{\cdot-}$. The $O_2^{\cdot-}$ in the chloroplast occurs when electrons are transferred directly from photosystem I to oxygen (Hossain and Asada, 1984; Causin et al., 2009; Roqueiro et al., 2012). Dissipation of excess energy absorbed by the photosynthetic apparatus is a fundamental process that is essential for the survival of photosynthetic organisms. It prevents the photo-oxidative damage that occurs when excited chlorophyll molecules improperly transfer their higher energy state to oxygen or neighboring molecules and convert them into reactive molecules or toxic radicals (Rizhsky et al., 2003). The ·OH is highly destructive to the lipids, nucleic acids and proteins (Morel y Barouki, 1999). However, $O_2^{\cdot-}$ and H_2O_2 are necessary for the lignification process and functions as signal for defense response to pathogen infection (Ruiz-Lozano, 2003). H_2O_2 is an important signaling compound when considering plant-pathogenic microorganisms, the rhizobial interaction and other processes related to the biology of root gravitropism and elongation (Fester and Hause, 2005).

The enzymes that detoxify free radicals are not uniformly distributed, are placed in cellular compartments (Ruiz-Lozano, 2003). As well as H_2O_2 is decomposed by catalase (CAT) in the cytosol, in the glyoxysomes and peroxisomes is detoxified by ascorbate peroxidase (APX) on cytosol and plastid turning it into H_2O; particularly APX decomposes through the cycle ascorbate - glutathione (ASC-GSH) (Mittler et al., 2004). It is a major antioxidant defenses in plants and their recognized role is the intervention in eliminating H_2O_2 in the chloroplast. However, all components of this pathway are also present in mitochondria and peroxisomes of leaf (Jimenez et al., 1997). The cycle involves the concerted action of four enzymes: APX, dehydroascorbate reductase (DHAR), monodehydroascorbate reductase (MDHAR) and glutathione reductase (GR), and requires a continuous supply of ascorbate, thiols and reduced pyridine nucleotide (Dalton et al., 1986, 1992). The first reaction of the pathway is catalyzed by APX. The APX peroxidase is the most important in the detoxification of H_2O_2 in chloroplasts (Noctor and Foyer, 1998; Iturbe-Ormaetxe et al., 2001; Ruiz-Lozano, 2003). Superoxide dismutase (SOD) catalyzes the dismutation of $O_2^{\cdot-}$ to O_2 and H_2O_2, then the CAT and other peroxidases remove H_2O_2 since H_2O_2 is a strong oxidant that rapidly oxidized thiol groups and therefore can not accumulate in excess (Ruiz-Lozano, 2003). Three classes of SOD's in plants depending on the cofactor used as the active site (Mn, Fe or Cu-Zn) are known. Thus, we find MnSOD's and FeSOD's structurally related but different location. MnSOD is found in mitochondria while the FeSOD is located in the chloroplast. The CuZnSOD is not structurally related to the above and can be found in the cytosol and in the chloroplast.

Mycorrhizae, Water and Oxidative Stress

Plants grow in association with a number of soil microorganisms capable of alleviating the stress symptoms (Ruiz-Lozano, 2003). The arbuscular mycorrhizal fungi (AMF) are associations between roots and mycorrhizal fungi grouped in Glomerales order (Schüβler et al., 2001). The AMF is not specific host strain since it can colonize different plant species, however, the ability of the different species to promote the growth of the same host is variable. Symbiosis plant - arbuscular mycorrhizae are present in all natural ecosystems, even those affected by adverse environmental conditions. It can be defined as a more efficient specialized system for nutrient uptake and transfer to the single root, on the other hand, the roots of mycorrhizal plants scan a much larger volume having larger areas of absorption that the roots of non-mycorrhizal plants (Tanu et al., 2005). Mycorrhizae also enhance adaptation to field conditions in soil with little water available (Waterer and Coltman, 1989; Subramanian and Charest, 1997; Pinior et al., 2005; Zhu et al., 2011), plant survival (Wilson et al., 1991; Stahl et al., 1998; Narayan et al., 2013) and transplant success is increased in mycorrhizal plants (Subhan et al., 1998; Espeleta et al., 1999; Bompadre et al., 2014).

Water stress modifies the metabolism of the plant and can affect the content of sugars, soluble proteins and amino acids. It was observed that the soluble sugars increase in AMF plants under water deficit conditions (Wu et al., 2006b). Auge (2001) added that the contribution of arbuscular mycorrhizal symbiosis to drought tolerance is a combination of physical, nutritional and physiological effects (Narayan et al., 2013). It is not known exactly the mechanism by which arbuscular mycorrhizae improve plant water relations. Potential explanations include extended water absorption through the external hyphae, stomatal regulation by hormonal signals, an indirect effect of the absorption of phosphorus via the water connection and an osmotic adjustment in mycorrhizal plants (Wu et al., 2006b; Narayan et al., 2013). Also in mycorrhizas can increase the intensity of lateral roots branching (Citernesi et al., 1998) (Figure 1).

Is growing existing information on antioxidant enzyme activities in arbuscular mycorrhizal symbiosis. Has focused on the role of SOD's, while the rest of the enzymes involved in the elimination of ROS such as CAT, APX or GR are in study (Ruiz-Lozano, 2003; Wu et al., 2006b; Zhu et al., 2011; Bompadre et al., 2013). Many authors found an increase in antioxidant enzymes against both biotic and abiotic stress. In turn, the AMF can reduce oxidative damage to lipids through the reduction in the level of malondialdehyde (MDA). The protection against oxidative damage to lipids has also been increasingly studied (Alguacil et al. 2003; Porcel and Ruiz-Lozano, 2004; Wu et al., 2006b; Zhu et al., 2011).

Figure 1. Different benefits provided by colonization with Rhizophagus irregularis in plants; shoot elongation of a, Sinningia speciosa and b, Saintpaulia ionantha; flowering advance in Viola cornuta c, before flowering and d, after flowering; survival improvement of e, Punica granatum; soil aggregation of f, Nephrolephis exaltata. Abbreviations: SM (control plants without mycorrhizal inoculation); CM (R. irregularis inoculation); SS (without soil addition); CS (with soil addition).

PROPAGATION BY CUTTINGS, NURSERY PRODUCTION

The term cutting involves removing a piece from a mother plant and that piece regenerates the lost tissues. It is an asexual propagation via cuttings from stems, leaves and roots commonly used to propagate some commercial ornamental and woody plants. Usually propagation by cuttings of shrubs are easy to root, however, the propagation of the species of trees are more difficult and require intensive care. In most cases the application of rooting hormones (e.g. indole butyric acid, IBA). Is necessary to maintain high humidity and to ensure successful rooting, one way is by the use of intermittent irrigation MIST driven by a clock. In turn, the cuttings should be protected from direct light intensity by covering with plastic or inside a nursery.

There are four types of cuttings based on the stage of the mother plant:

Herbaceous cuttings: They are made from non-woody herbaceous plants such as chrysanthemums and dahlias.

Non-woody cuttings: Prepared from woody plants begin to mature (new growth, soft stems) are the first branches from a mother plant within the first year of growth that have not reached maturity. In this case, soft leaves are very sensitive and must take special care to avoid dehydration.

Semi-woody cuttings: from partially mature wood in the growing season after flowering. Wood in general is reasonably firm and the leaves have a mature size. Some conifers and evergreen plants are propagated in this group. In this group, olive and pomegranate garden spread.

Woody cuttings: From dormant stems in winter or early spring. Plants generally are completely dormant or without signs of active growth. Wood is strong and not easily distorted.

When new species or varieties begin to be grown on a commercial scale, one problem that arises is the provision of quality plants. Although there is no exact definition of plant quality and the same observer subjectivity involved, you can point out certain quantitative parameters, besides the aesthetic (qualitative) that, even if not standardized, are perceived by the consumer and are positively related with a superior product: plant form, number and size of branches, stem- root proper relationship, high growth rate, especially in the early stages to later rustification implantation in place definitive (Divo de Sesar et al., 2003). While the initial guideline to develop and select a protocol on a commercial scale is rooting percentage obtained, the final criteria were based on the seedlings to survive and grow (Liptay et al., 1998; Divo de Sesar et al., 2003).

Preferably from the top of the mother plant so as to obtain healthy and disease free individuals without flowers so that the energy is used for growth of roots cuttings are chosen. The fertility status of the mother plant influence the rooting of cuttings. Mother plants with symptoms of nutritional deficiency are discarded. Cuttings from lateral branches of mother plants rooted better than those from terminal branches (Bryant, 1995; Hartmann et al., 1996).

The length of the cuttings range from 4 to 6 cm, the cutting is performed below node and allowed at least one internode with intact leaves and buds, the leaves of the bottom of the cutting are removed. For species which rooted troubles usually a diagonal incision is made at the edge of the cutting to increase rooting area and allow greater contact with rooting hormone. The hormone can be prepared in different concentrations adsorbed to an inert carrier (industrial talc plaster free) according to the difficulty of rooting of the cutting, these values range from 2500 ppm to 4000 ppm (Divo de Sesar et al., 2003; Knight et al., 2005).

The structures developed for use in systems productive propagation of plants are designed to maintain an environment with low pressure deficit vapor, minimizing lost by transpiration, acceptable temperatures for processes of regeneration that will occur at the base of the cuttings avoiding heat stress of the sheets, and light levels suitable for photosynthesis and photo-assimilated production that will be used in the maintenance of the cutting and the root growth. The substrate to place the cuttings must be sterile, low in fertility and well drained to provide good aeration, the most widely used substrate is perlite. Once the carrier was moistened it proceeds to introduce a third part in depth vertically, the spacing between cuttings varies by species but should be long enough to allow all the leaves are in contact with sunlight. Rooting time varies depending on the species used, environmental conditions and the type of cut made. After this period the cutting is "rusticated" and is ready to be transplanted for the first time (Bryant, 1995; Hartmann et al., 1996; Divo de Sesar et al., 2003) (Figure 2).

Transplanting, a horticultural process as old as civilization, is generally misunderstood and often results in the loss of many beautiful or valuable plants. Even under the most optimum conditions, the first year is full of great hazards for the newly transplanted specimen (Koller, 1977). That is why at transplanting the roots formed are not fully functional. Since these roots hairs aso secrete substances helping the movement of the root into new soil areas, their regeneration is extremely important.

Figure 2. Cutting management a, cutting propagation on a raised rooting table under nursery; b, olive cuttings rooted for a period of 60 days ready to be transplanted.

It commits the following steps and if not kept in controlled environments (shaded and wet) a delay is seen in rustification and in the subsequent growth increasing time in nursery (Lagoutte et al., 2009).

Applicability of Mycorrhizae and Mycorrhizal Production Systems

Current technologies are based on agricultural production systems that emphasize the study and management of microbial interactions in the soil-plant interface. In the Mediterranean there are new technologies for the production of olive trees (*Olea europaea* L.) based on the application of microbial inoculants as bioprotectors, fitoestimuladores or biofertilizers for greenhouse production of quality seedlings. The management of mycorrhizal associations have been proposed for the role of showing the same in terms of plant growth and plant health (Calvente et al., 2004; Porras Piedra et al., 2005; Isfendiyaroglu et al., 2009).

A critical step in the implementation of the AMF technology is the appropriate selection of effective fungal isolates for use as inoculants. For this purpose it is recommended to test effective native ecotipes (Calvente et al., 2004; Marin, 2005) to allow take advantage of the numerous benefits granted (Gianinazzi et al., 2010).

Modern farming systems, particularly those carried out in greenhouses (seedlings, cuttings, micro-propagated plants) include the incorporation of large amounts of soluble fertilizers to be sustainables (Marin, 2005) from which only a part is used by plants. Most soils contain substantial reserves of total phosphorus, but remains relatively inert and less than 10% enters the plant-animal cycle (Sundara et al., 2002). Plants acquire P in the soil solution as phosphate anion; mainly HPO_4^{2-} and $H_2PO_4^-$. However, the phosphate anions are highly reactive and can be immobilized via its precipitation type cation Ca^{2+}, Mg^{2+}, Fe^{3+} and Al^{3+}, or adsorbed on oxides of iron and aluminum, and calcium silicates and aluminum depending of the particular properties of the soil (Wakelin et al., 2004; Narayan et al., 2013). Thus, in acid soils, the reaction products are Fe-P and Al-P and Ca-P in calcareous soils (Sundara et al., 2002). The addition of these organisms makes the use of these compounds better reducing production costs (Barea and Jeffries, 1995).

Many plants known for transplant difficulty are lost because the proper soil organisms are lacking (Koller, 1977). In the management of mycorrhizal inoculants rhizospheric organisms can be successfully applied at different times depending on the type of production plant being particularly important in the early stages of growth. It is advisable to use:

Seedling production: in this technique seeds are used, arbuscular mycorrhizal fungi are incorporated at the time of transplantation since the early stages of seedling growth depend on the nutrients of the seeds. However, the advantage of including AMF in this step has the advantage of using small amount of inoculum (Marin et al., 2002).

Micropropagated plants: in this technique the beneficial microorganisms are absent, leaving the plants exposed to various types of stress from the first steps of growth. The addition of mycorrhizal inoculum in the early stages of acclimatization makes this practice more successful (Bompadre, 2002; Pérgola, 2002; Pinior et al., 2005).

Production of cuttings: the use of mycorrhizal fungi inoculum is growing, they promote the formation of lateral roots (Citernesi et al., 1998; Scagel et al., 2003) as growth regulators such as auxins and cytokines are synthesized (Gianinazzi, 1991; Raman and Selvaraj, 2005) may therefore be applied from the first stages of early rooting achieving an early inoculation allowing greater benefits as soon as possible.

In turn, the benefits resulting from the use of AMF are divided into two conditions: in the greenhouse effect and effect outside it. The beneficial effects of the symbiosis in practice performed in the greenhouse may be less apparent due to the controlled conditions are achieved within the same. Mycorrhizae probably not accelerate the process of growing in the greenhouse but promote success transplanting outside (Marin, 2005). So a good root colonization with appropriate mycorrhiza attenuate the effects of dehydration (drought and temperature) experienced by plants after transplanting during the first stages in the soil. As describes Gianinazzi et al. (2010) most of the research has been carried out under controlled conditions at early stages of plant growth but a few studies conducted in the field or in the greenhouse under real production conditions confirm these results.

Mycorrhizae and Its Benefit against Oxidative Stress in Handling Cuttings

The cutting propagation practice has various stressing moments. After rooting stage a first transplant is needed, as the subsequent adaptation to a new substrate condition. Even in the best transplant operation, the root system is greatly reduced. On woody plants this loss must be compensated by removing at least one-third of the growing vegetative shoots. If this essential step is neglected it will intensify the following plant reactions: when vegetative buds burst into growth, the plant immediately undergoes moisture stress as there are not enough roots to support the lush growth (Koller, 1977).

In a previous study about the use of commercial mycorrhiza under nursery or landscape conditions Carpio et al. (2003) observed a higher survival of AMF plants in landscape condition. Thus the initial benefit of AMF inoculation should come during transplant establishment and growth. In a transplant evaluation study of *Olea europaea* cv. Manzanilla plants (olive) inoculated with two different strains of *Rhizophagus irregularis,* it was found that when they were subjected to transplant, before AM inoculation, olive plant growth was affected by transplant; however after a year of inoculation, at the second transplant instance, growth was improved by the co-inoculation of both strains (Bompadre et al., 2014). Several authors have reported a denser root system in small mycorrhizal plants (Davies et al., 2002; Alguacil et al., 2003; Franco et al., 2011). Calvente et al. (2004) found a growing improvement of Arbequina and Leccino olive cultivar cuttings with selected AMF species *R. irregularis* and *Glomus viscosum.* Also Porras-Soriano et al. (2009) observed higher effectiveness of Cornicabra olive cultivar plants inoculated with *Funneliformis mosseae* compared to *R. irregularis* and *Claroideoglomus claroideum* against saline stress. The capability of AMF to protect plants from the detrimental effects of salt stress may depend on the behaviour of each species. Alguacil et al. (2003) found different levels of effectiveness in improving the growth of *O. europaea*, *Retama sphaerocarpa*, and *Rhamnus lycioides* with a mixture of native AMF being equal to or more effective than an allochthonous AMF.

Olive trees are able to up-regulate the enzymatic antioxidant system. Several authors found a decrease in MDA on woody plants (Bacelar et al., 2006; Wu et al., 2006b; Bompadre et al., 2014), also founding different levels of susceptibility to lipid peroxidation. Ruíz-Lozano et al. (2012) observed that mycorrhizal plants possess an enhanced activity of several antioxidant enzymes, but it has been shown that the response of the individual enzyme depends on the fungal and host plant species.

Under nursery growing condition, plants suffer different drought levels during the day. An improvement in the survival rate of AM inoculated plants under water stress conditions have been previously observed (Alguacil et al., 2003; Porcel et al., 2003; Al-Karaki et al., 2004; Wu et al., 2008; Bompadre et al., 2013). Other authors have also found differences related to the ability of different AM strains to relieve water stress. For example, *Funneliformis monosporus* isolates have proved to be less effective than *F. mosseae* on wheat plants in water deficit condition (Al-Karaki et al., 1998). However, olive plants inoculated with two strains of *R. irregularis* (with different growing behavior), showed that co-inoculation and one single strain inoculation were more efficient under low water conditions (Bompadre et al., 2013).

Several authors have associated biomass and nutrient status with increased activities of antioxidant enzymes in mycorrhizal plants (Alguacil et al., 2003; Huang et al., 2008; Roldán et al., 2008; Franco et al., 2011). When *R. irregularis* strains were inoculated in *Punica granatum* plants, under nursery conditions, it was shown that early AMF inoculation during pomegranate propagation protects plants against abiotic stress (Bompadre et al., 2014b). The response to oxidative stress depends on many factors, including the tissue under analysis and the extent of the stress. In pomegranate plants co-inoculation favor the antioxidative response. On the other hand, Wu et al. (2006b) found that AMF plants show higher antioxidant enzymes during adverse conditions, and the breakage of ROS is alleviated by AMF colonization.

Conclusion

Under nursery practices there is several stressing factors during cutting development that causes delay in plant development or important lose of valuable plants. A good management of different strains of arbuscular mycorrhizal fungi can reverse the stress condition. Inoculation with two AM fungi does not always bring additional benefit to the host plant in comparison to single inoculation. However, an early mycorrhizal inoculation increases plant growth and protects against oxidative stress improving transplant stress tolerance.

Acknowledgments

We thank UNQ, UBA, CONICET and Agencia Nacional de Promoción Científica y Tecnológica (ANPCyT) for financial support.

References

A. Porras-Soriano, M. L. Soriano-Martín, A. Porras-Piedra, and R. Azcón, "Arbuscular mycorrhizal fungi increased growth, nutrient uptake and tolerance to salinity in olive trees under nursery conditions," *Journal of Plant Physiology*, vol. 166, no. 13, pp. 1350-1359, 2009.

Alguacil, MM; Hernández, JA; Caravaca, F; Portillo, B; Roldán, A. Antioxidant enzyme activities in shoots from three mycorrhizal shrub species afforested in a degraded semi-arid soil. *Physiologiae Plantarum,* 2003 118, 562-570.

Al-Karaki G, McMichael B, Zak J (2004) Field response of wheat to arbuscular mycorrhizal fungi and drought stress. Mycorriza 14:263-269.

Al-Karaki GN (1998) Benefit, cost and water-use efficiency of arbuscular mycorrhizal durum wheat grown under drought stress. Mycorriza 8(1):41-45.

Asada, K. The water-water cycle in cloroplasts: scavenging of active oxigen and dissipation of excess photons. *Annual Review of Plant Physiology and Plant Molecular Biology,* 1999 50, 601-639.

Asada, K; Takahashi, M. Production and scavenging of active oxygen in photosynthesis. In: Kye DJ; Osmond CB; Arntzen CJ (eds.)

Photoinhibition, Topics in Photosynthesis. Amsterdam: Elsevier; 1987; 227-287.

Augé, RM. Water relations, drought and vesicular arbuscular symbiosis. *Mycorrhiza,* 2001 11, 3-42.

Barea J.M. y Jeffries P. (1995). Arbuscular mycorrhizas in sustainable soil plant systems. En: Mycorrhiza Structure Function, Molecular Biology and Biotechnology, Hock, B. and Varma, A. Eds. Springer-Verlag, Heidelberg. pp. 521-559.

Bompadre M.J. (2002). Dos metodologías de inoculación con *Glomus intraradices* y *Gigaspora rosea* en vástagos micropropagados de helecho (*Nephrolepis exaltata* cultivar Boston). Tesis de Licenciatura.

Bompadre MJ, Pérgola M, Fernández Bidondo L, Colombo RP, Silvani VA, Pardo AG, Ocampo JA, Godeas AM 2014. Evaluation of arbuscular mycorrhizal fungi capacity to alleviate abiotic stress of olive (*Olea europaea* L.) plants at different transplant conditions. The Scientific World Journal. 1-12.

Bompadre MJ, Silvani VA, Fernández Bidondo L, Ríos de Molina MDC, Colombo RP, Pardo AG, Godeas AM 2014b. Arbuscular mycorrhizal fungi alleviate oxidative stress in pomegranate plants growing under different irrigation conditions. Botany. 92(3): 187-193.

Bray E.A. (1997). Plant responses to water deficit. *Trends in Plant Science.* 2(2): 48-54.

Bray E.A., Bailey-Serres J., Weretilnyk E. (2000). Responses to Abiotic Stresses. En: Biochemistry and molecular biology of plants. Buchanan B., Gruissem R., Jones R. Eds. American society of plant physiologists. pp. 1158-1203.

Bryant G. (1995). Propagation Handbook. Stackpole Books: Mechanicsburg, Pennsylvania.

Calvente R., Cano C., Ferrol N., Azcon-Aguilar C., Barea J.M. (2004). Analysing natural diversity of arbuscular mycorrhizal fungi in olive tree (*Olea europaea* L.) plantations and assessment of the effectiveness of native fungal isolates as inoculants for commercial cultivars of olive plantlets. *Applied Soil Ecology.* 26: 11-19.

Carpio, L.A., Davies, Jr., F.T., Arnold, M.A., 2005. Arbuscular mycorrhizal fungi, organic and inorganic controlled-release fertilizers: effect on growth and leachate of container-grown bush morning glory (*Ipomoea carnea* ssp. *fistulosa*) under high production temperatures. J Amer Soc Hort Sci 130(1), 131-139.

Causin, H.F., Roberts, I.N., Criado, M.V., Gallego, S.M., Pena, L.B., Ríos, M.C., Barneix, A.J. 2009. Changes in hydrogen peroxide homeostasis and cytokinin levels contribute to the regulation of shade-induced senescence in wheat leaves. J. Plant Sci. 177: 698-704.

Citernesi, A.S., Vitagliano, C, Giovannetti, M. (1998). Plant growth and root system morphology of *Olea europaea L.* Rooted cuttings as influenced by arbuscular mycorrhizas. *Journal of Horticultural Science Biotechnology*. 73(5): 647-654.

Dalton D.A. (1995). Antioxidant defenses of plants and fungi. En: Oxidative stress and antioxidant defenses in biology. Ahmad S. Ed. Chapman and Hall, New York. pp. 298-355.

Dalton D.A., Langeberg L., Robbins M. (1992). Purification and characterization of monodehidroascorbate reductase from soybean root nodules. *Archives of Biochemistry and Biophysics*. 292: 281-286.

Dalton D.A., Rusell S.A., Hanus F.J., Pascoe G.A., Evans H.J. (1986). Enzymatic reactions of ascorbate and glutathione that prevent peroxide damage in soybean root nodules. *Proeedings in Natural Academy of Science* USA. 83: 3811-3815.

Davies, F.T. Jr., Olade-Portugal, V., Aguilera-Gomez, L., Alvarado, M.J., Ferrera-Cerrato, R.C., Boutton, T.W. 2002. Alleviation of drought stress of Chile ancho pepper (*Capsicum annuum* L. cv. San Luis) with arbuscular mycorrhiza indigenous to Mexico. Scientia Hort. 92: 347-359.

Del Rio L.A., Sandalio L.M., Palma J.M., Bueno P., Corpas F.J. (1992). Metabolism of oxygen radicals in peroxisomes and cellular implications. *Free Radicals on Biology and Medicine.* 13: 557-580.

Divo de Sesar M, D'Ambrogio A, Boquete J, Vilella F, Stella A (2003). Determinación del momento óptimo de aplicación de 6-bencilaminopurina y correspondencia con la secuencia histológica en el proceso de enraizamiento de estacas semileñosas de *Jasminum mensyi* Han-ce (Oleaceae). In: A.J. Paséale (ed.), Floricultura en la Argentina, Investigación y Tecnología de Producción. Editorial Facultad de Agronomía, Universidad de Buenos Aires, Argentina. Pp 27-36, 468 pp. ISBN: 978-950-29-0746-8.

E. A. Bacelar, D. L. Santos, J. M. Moutinho-Pereira, B. C. Gonçalves, H. F. Ferreira, and C. M. Correia, "Inmediate responses and adaptative strategies of three olive cultivars under contrasting water availability regimes: Changes on structure and chemical composition of foliage and oxidative damage," *Plant Science*, vol. 170, pp. 596-605, 2006.

Espeleta J.F., Eissentat D.M., Graham J.H. (1999). Citrus root responses to localized drying soil: a new approach to studying mycorrhizal effects on the roots of mature trees. *Plant Soil.* 206: 1-10.

Fester T. y Hause G. (2005). Accumulation of reactive oxigen species in arbuscular mycorrhizal roots. *Mycorrhiza.* 15: 373-379.

Franco, J.A., Bañón, S., Vicente, M.J., Miralles, J., Martínez-Sánchez, J.J. 2011. Root development in horticultural plants grown under abiotic stress conditions – a review. J Hort. Sci. & Biotech. 86(6): 543-556.

Gianinazzi S. (1991). Vesicular-Arbuscular (endo)-mycorrhizas: Cellular, biochemical and genetics mycorrhizal roots. *Plant and Soil.* 71: 197-209.

Hartmann H.T., Kester D.E., Davies F.T., Geneve R.L. (1996). Plant Propagation, Principles and Practices. Prentice Hall: Upper Saddle River, New Jersey.

Hossain M.A. y Asada K. (1984). Inactivation of ascorbate peroxidase in spinach chloroplasts on dark addition of hydrogen peroxide: its protection by ascorbate. *Plant Cell Physiology.* 25: 1285-1295.

Huang, L.L., Yang, C., Zhao, Y., Xu, X., Xu, Q., Li, G.Z., Cao, J., Herbert, S.J. 2008. Antioxidant defenses of mycorrhizal fungus infection against SO_2- induced oxidative stress in *Avena nuda* seedlings. Bull. Environ. Cont. Toxicol. 81: 440-444.

Isfendiyaroglu M, Ozeker E, Baser S (2009) Rooting of 'Ayvalik' olive cuttings in different media. Spanish J of Agricultural Research. 7(1): 165-172.

Iturbe-Ormaetxe I., Matamoros M.A., Rubio M.C., Dalton D.A., Becana M. (2001). The antioxidants of legume nodule mitochondria. *Molecular Plant Microbe International.* 14: 1189-1196.

J. M. Ruíz-Lozano, R. Porcel, C. Azcón, and R. Aroca, "Regulation by arbuscular mycorrhizae of the integrated physiological response to salinity in plants: new challenges in physiological and molecular studies," *Journal of Experimental Botany*, vol. 63, no. 11, pp. 4033-4044, 2012.

Jimenez A., Hernandez J.A., Del Rio L.A., Sevilla F. (1997). Evidence for the presence of the ascorbate-glutathione cycle in mithocondria and peroxisomes of pea leaves. *Plant Physiologist.* 114: 275-284.

Koller G.L., 1977. Transplanting stress – a view from the plant's perspective. Arnoldia, *37*(5): 230-241.

Lagoutte S, Divo de Sesar M, Vilella F (2009). Efecto del tamaño de celdas y citoquininas en el crecimiento de plantas de petunia. Phyton 78(1): 31-36.

Liptay A, Sikkema P, Fonteno W (1998). Transplant growth control through water deficit stress-A review. HortTechnol 8(4): 1-5.

M. J. Bompadre, M. C. Rios De Molina, R. P. Colombo, L. Fernandez Bidondo, V. A. Silvani, A. G. Pardo, J. A. Ocampo, and A. M. Godeas, (2013) "Differential efficiency of two strains of the arbuscular mycorrhizal fungus *Rhizophagus irregularis* on olive (*Olea europaea*) plants under two water regimes," *Symbiosis,* 61: 105-112.

Marin M. (2005) Arbuscular mycorrhizal inoculation in nursery practice. En: Handbook of Microbial Biofertilizers. Food Products Press® An Imprint of the Haworth Press, Inc New York. London. Oxford Rai M.K. Ed. pp. 289-324.

Miller G, Suzuki N, Ciftci-Yilmaz S, Mittler R 2010. Reactive oxygen species homeostasis and signalling Turing drought and salinity stresses. Plant, Cell and Environment 33, 453-467.

Mittler R 2002 Oxidative stress, antioxidants and stress tolerante. Trends in Plant Science 7(9), 405-410.

Mittler R., Vanderauwera S., Gollery M., van Breusegem F., 2004. Reactive oxygen gene network of plants. Trends in Plant Science, *9*(10): 490-498.

Morel Y. y Barouki R. (1999). Repression of gene expression by oxidative stress. *Biochemical Society.* 342: 481-496.

Narayan RP, Khare V, Kehri HK 2013. Role of AM fungi in reclamation of salt affected soils: a review. International journal of Biology, Pharmacy and Allied Sciences. 2(5): 1167-1187.

Nilsen E.T. y Orcutt D.M. (1996). The physiology of plants under stress. John Wiley & Sons Inc., Nueva York.

Noctor G. y Foyer C.H. (1998). Ascorbate and glutathione: Keeping active oxygen under control. *Annual Review of Plant Physiology and Plant Molecular Biology.* 49: 249-279.

P. Knight, C. H. Coker, J. M. Anderson, D. S. Murchison, C. E. Watson, "Mist interval and K-IBA concentration influence rooting of orange and mountain azalea," *Native Plants*, pp. 111-117, 2005.

Pérgola M. (2002). Inoculación con *Glomus intraradices* en vástagos microporpagados de diferentes especies ornamentales. Tesis de Licenciatura.

Pinior A, Grunewaldt-Stöcker G, von Alten H, Strasser RJ (2005) Mycorrhizal impacto n drought stress tolerante of rose plants probed by chlorophyll *a* fluorescente, proline content and visual scoring. Mycorrhiza 15: 596-605.

Porcel R, Barea JM, Ruíz-Lozano JM (2003) Antioxidant activities in mycorrhizal soybean plants under drought stress and their possible relationship to the process of nodule senescence. New Phytol 157:135-143.

Porcel, R., Ruíz-Lozano, J.M. 2004. Arbuscular mycorrhizal influence on leaf water potential, solute accumulation, and oxidative stress in soybean plants subjected to drought stress. J. Exp. Bot. 55(403): 1743-1750.

Porras Piedra A., Soriano Martín M.L., Porras Soriano A., Fernández Izquierdo G. (2005). Influence of arbuscular mycorrhizas on the growth rate of mist-propagated olive plantlets. *Spanish Journal of Agricultural Research.* 3(1): 98-105.

Raman N. y Selvaraj T. (2005). Tripartite relationship of Rhizobium, AMF, and host in growth promotion. En Handbook of microbial Biofertilizers. Food Products Press®. An Imprint of the Haworth Press, Inc New York. London. Oxford Rai M.K. Ed. pp. 51-88.

Rizhsky, L., Liang, H., Mittler, R. 2003. The water-water cycle is essential for chloroplast protection in the absence of stress. J. Biol. Chem. 278(40): 38921–38925.

Roldán, A., Díaz-Vivancos, P., Hernández, J.A., Carrasco, L., Caravaca, F. 2008. Superoxide dismutase and total peroxidase activities in relation to drought recovery performance on mycorrhizal shrub seedlings grown in an amended semiarid soil. J. Plant Physiol. 165: 715-722.

Roqueiro, G., Maldonado, S., Ríos, (de Molina) M.C., Maroder, H. 2012. Fluctuation of oxidative stress indicators in *Salix nigra* seeds during priming. J. Exper. Bot. 63(10): 3631-3642.

Ruiz-Lozano J.M. (2003). Arbuscular mycorrhizal symbiosis and alleviation of osmotic stress. New perspectives for molecular studies. *Mycorrhiza.* 13: 309-317.

S. Gianinazzi, A. Gollotte, M. N. Binet, D. van Tuinen, D. Redecker, and D. Wipf, "Agroecology: the key role of arbuscular mycorrhizas in ecosystem services," *Mycorrhiza,* vol. 20, pp. 519-530, 2010.

Sanchez-Diaz M. y Aguirreolea J. (2002). El agua en la planta. Fundamentos de fisiología vegetal. Azcon-Bieto J. y Talon M., Eds. McGraw-Hill-Interamericana. pp. 17-30.

Scagel CF, Keddy K, Armstrong JM (2003) Mycorrhizal fungi in rooting substrate influences the quantity and quality of roots on stem cuttings of hick's yew. HortTechnology. 13(1): 62-66.

Scandalios J.G., Guan L., Polidoros A. N. (1997). Catalases in plants: Gene structure, properties, regulation and expression. En: Oxidative stress and the molecular biology of antioxidants defenses. Scandalios J.G. Eds. Cold springer harbor, New York. pp. 343-406.

Schüβler A., Schwarzott D., Walker C. (2001). A new fungal phylum, the *Glomeromycota*: phylogeny and evolution. *Mycological Research.* 105(12): 1413-1421.

Smirnoff N. (1998). Plant resistance to environmental stress. *Current Opinion in Biotechnology*. 9: 214-219.

Stahl P.D., Schuman G.E., Frost S.M., Williams S.E. (1998). Arbuscular Mycorrhizal and water stress tolerance of Wyoming big sagebrush seedling *Soil Science in Society of American Journal*. 62: 1309-1313.

Subhan S., Sharmila P., Pardha Saradhi P. (1998). *Glomus fasciculatum* alleviates transplantation shock of micropropagated *Sesbania sesban. Plant Cell Repports.* 17: 268-272.

Subramanian K.S., y Charest C. (1997). Nutritional, growth and reproductive responses of maize during and after drought stress at tasselling. *Mycorrhiza.* 7: 25-32.

Sundara B., Natarajan V., Hari K. (2002). Influence of phosphorous solubilizing bacteria on the changes in soil available phosphorous and sugarcane and sugar yields. *Field Crops Research.* 77: 43-49.

Tanu, Prakash A., Adholeya A. (2005). Potential of arbuscular micorrhizae in organic farming systems. En: Handbook of Microbial Biofertilizers. Food Products Press® An Imprint of the Haworth Press, Inc New York. London. Oxford Rai M.K. Ed. pp. 223-239.

Wakelin S. A., Warren R. A., Harvey P. R., Ryder M. H. (2004). Phosphate solubilization by *Penicillium* spp. Closely associated with wheat roots. *Biology of Fertil Soils.* 40: 36-43.

Wang W.X., Vinocur B., Shoseyov O., Altman A. (2001). Biotechnology of plant osmotic stress tolerance: physiological and molecular considerations. *Acta Horticultural.* 560: 285-292.

Waterer D.R. y Coltman R.R. (1989). Response of mycorrhizal bell peppers to inoculation timing, phosphorus and water stress. *HortScience.* 24: 688-690.

Wilson J, Munro RC, Ingleby K, Mason PA, Jefwa J, Muthoka PN, Dick JM Leakey RR. (1991). Tree establishment in semiarid lands of Kenya-role of mycorrhizal inoculation and water –retaining polymer. *Forest Ecology and Management*. 45: 153-163.

Wu Q.S., Zou Y.N., Xia R.X. (2006b). Effects of water stress and arbuscular mycorrhizal fungi on reactive oxigen metabolism and antioxidant production by citrus (*Citrus tangerine*) roots. *European Journal of Soil Biology*. 42: 166-172.

Wu QS, Xia RX, Zou YN (2008) Improved soil structure and citrus growth after inoculation with three arbuscular mycorrhizal fungi under drought stress. Europ J Soil Biol 44:122-128.

Zhu, X., Song, F., Liu, S. 2011. Arbuscular mycorrhiza impacts on drought stress of maize plants by lipid peroxidation, proline content and activity of antioxidant system. J. Food Agric. & Environ. 9(2): 583-587.

In: Mycorrhizas
Editor: Elaine Warwick
ISBN: 978-1-63321-436-1

Chapter 6

VESICULAR-ARBUSCULAR MYCORRHIZAE IN WASTEWATER IRRIGATED AGRICULTURE

Gabriel Camarena Gutiérrez[*]
Unidad de Morfología y Función. FES Iztacala,
Universidad Nacional Autónoma de México. Tlalnepantla, México

ABSTRACT

This article reviews the state of the art with respect to the environmental effects of wastewater irrigated agriculture on arbuscular mycorrhizal fungi in arid and semi-arid regions on a field scale. Petroleum, tannery and pharmaceutical compounds emerge as pollutants when wastewater is used for irrigation without any previous treatment, situation which is now the object of new studies. In the last few decades, the amounts of detergents, fats, oils, heavy metals and synthesized organic compounds generated by industrial activities have increased, and some 10,000 new organic compounds are added each year. and released into the environment for direct or indirect application over a long period of time. In both developed and developing countries, the most prevalent practice is the application of municipal wastewater (both treated and untreated) to land. Impact from wastewater on agricultural soil, is mainly due to the presence of high nutrient contents (Nitrogen and Phosphorus), high total dissolved solids and compounds pollutants, which are added to the soil over time. Arbuscular mycorrhizal plants have two pathways that

[*] Email: datura@unam.mx.

can contribute to nutrient uptake from soil; direct uptake from the rhizosphere by root epidermal cells and the vesicular arbuscular mycorrhyzae uptake pathway via the fungi. The establishment of the vesicular arbuscular mycorrhizae symbiosis can be disturbed by environmental conditions, including phosphorus (P) availability which can inhibit the symbiotic interaction. Interestingly, recent studies have demonstrated a cross-talk between phosphorus and nitrogen availabilities to control arbuscular mycorrhizal associations, thus indicating a high level of integration of mycorrhizal responses to mineral nutrition. Some authors have reported a direct effect of P on arbuscular mycorrhizal spore germination and hyphal growth. Evidence has also been gathered for downregulation by P of well established vesicular arbuscular mycorrhizae interactions, via a reduced production of root compounds. Arbuscular mycorrhizal symbiosis could be arrested almost completely by a high P supply at a very early stage, prior to the attachment of the fungus on the root. The synthesis and exudation of strigolactones are negatively affected by a high availability of P. Under conditions of P sufficiency, a strong reduction of strigolactone release would prevent the stimulation of AM fungi and hence the establishment of the interaction. The impact of arbuscular mycorrhizas, is starting to draw attention in the context of wastewater irrigation and phytoremediation of organics, and recent results have shown that these symbioses may be crucial for both plant establishment on contaminated sites.

INTRODUCTION

Agriculture is the primary supplier of food and fibre for humankind and maintaining its sustainability is, thus, of vital importance. Currently, however, many agricultural activities cannot be considered sustainable. Three-quarters of the irrigated area (192 million hectares) is located in developing countries (United Nations 2003). Frequently in these countries, wastewater is used to irrigate land. It is also estimated that one-tenth or more of the world's population consumes crops irrigated with wastewater. Due to diversity in industrial growth there is a considerable variation in the quantity and quality of wastewater generated from industry to industry. The volume of wastewater generated from pulp and paper mill varies from 227-455 litres per kg of paper produced. The value of spent wash generated during the process of alcohol production varied from 15-20 litre per litre of alcohol produced, on the other hand, the wastewater generated by a dairy industries being 2-8 litres per litres of milk processed. In textile industry up to 60 to 70% of all dyes applied are from azo dye group (-N=N). The release of azo dyes into the environment is a concern due to coloration of natural water bodies and toxicity to the plants and animals. Siemens et al. (2008) already documented that wastewater contains a wide range of pharmaceuticals. Reports on the occurrence of pharmaceuticals in wastewater-irrigated soils in Colorado, USA, Germany, China, and the metropolitan area of Paris, France illustrate that the contamination of soils with wastewater-derived pharmaceuticals is not limited to Mexico, but a global phenomenon. Several investigations have studied the occurrence of antibiotic resistance genes in different environmental compartments including wastewater, wastewater lagoons, surface waters, river sediments, pristine soils, and manured soils (Gatica J. and Cytryn E. 2013). Long-term irrigation of soils with untreated wastewater in the Mezquital Valley led to an accumulation of sulfamethoxazole, ciprofloxacin, and carbamazepine (Dalkman et al., 2012). Many of these compounds are now found in the wastewater from most municipalities and communities. (Khan, et al., 2009). Coke plant wastewater contains mass toxic organic pollutants, including phenolic compounds, thiocyanate, free and complexed cyanides, polynuclear aromatic hydrocarbons, aliphatic and heterocyclic compounds. The results suggest that coke plant wastewater can cause oxidative damage in roots and leaves of Vicia faba and root enzymes seemed more sensitive to the wastewater (Liu and Lv. 2011). In the developing countries heavy metal pollution has become serious due to mining, mineral, smelting and tannery industry (Wang et al. 2007). The leather industry is recognized as a serious environmental threat all over the world because it is associated with the generation of huge amounts of heavy metal. Tannery wastewater is being contaminated with high levels of metals (Fe, Cr, Zn, Mn, Cu). León City is Mexico's most important shoe-manufacturing area, with about 600 tanneries. More than 80% of them use chromium as tanning agent. Their wastewater go directly to the municipal drainage system, wich flows to the León river. Finally, these waters are used for irrigation. Armienta et al. (2001) reported high concentration of Cr in the soil and plants irrigated with tannery wastewater in the Leon Valley, Mexico. Chromium was mostly retained by the roots in all the plants and no correlation seems to exist between total Cr content in the soils and plants. The plants grow in polluted area under long-term exposures accumulate metals at high concentrations, which become toxic for their growth and metabolism. The amounts of detergents, fats, oils, heavy metals and synthesized organic compounds generated by industrial activities have increased, and some 10000 new organic compounds are added each year. Management of such effluents to avoid damage to the environment warrants urgent attention.

Wastewater Irrigation

In large cities of developing countries, food demand is partially supplied by urban agriculture performed by smallholders envolved in market-oriented farming. Although considered as an informal economy, urban agriculture generates significant incomes for the underprivileged part of the population (Sou-Dakouré 2013). In the city of Kumasi Ghana, around 11,900 ha of peri-urban agriculture areas use untreated wastewater, (Keraita et al., 2002). Pakistan estimated 30,600 ha (Ensink et al., 2004). Along the Musi River in India, an estimated 9675 ha of agricultural land is irrigated by wastewater (Ahmed et al., 2006). In the near eastern region, which is one of the most water scarce regions in the world, documented evidence shows that wastewater irrigation is extensive notably in Saudi Arabia, Kuwait, Israel, Jordan, Tunisia and Morocco (AQUASTAT– FAO database, 1997). In countries of the former Soviet Union, literature indicates that at least 70,100 ha are irrigated with wastewater (AQUASTAT– FAO database, 1997). Extensive use has also been documented in water scarce developed regions of Northern America, Europe and Australia.

In Latin America a total of more than 500,000 ha of agricultural land is irrigated with wastewater. Of these 350,000 ha are in Mexico alone (Peasey et al., 2000). Mexico City Metropolitan Zone wastewater is composed of a mixture of domestic, municipal, and industrial wastewater, and stormwater runoff. It receives no conventional treatment (Siebe, 1998; Jiménez, 1995) and is subject to mixing and natural transformation processes during its transport from the México City Metropolitan Zone to the irrigation region and further biodegradation and sedimentation in a storage reservoir. Approximately 55% of the nation's industry is located in the México City Metropolitan Zone (paper, food, chemicals, textiles, and automotive), with approximately 43% of the wastewater in the irrigation district of industrial origin, and 57% of domestic and municipal origin (5). Similar practices take place at a smaller scale in the cities of Guanajuato, Tlaxcala, Puebla, Cuernavaca, Toluca and Pachuca. The environmental impact of the wastewater reuse at these places has been recently recognized as a major problem that needs a multidisciplinary approach for its evaluation (Dominguez et al., 2004). Wastewater irrigation in the Mezquital Valley in Mexico provides 2,400 kg organic matter, 195 kg nitrogen, and 81 kg phosphorus ha^{-1} yr^{-1}, contributing to significant increases in crop yields (Jimenez, 2005).

The amount of nutrients applied via wastewater irrigation in the World can vary considerably if the wastewater is raw, treated or diluted with stream water, constituent loads, frequency and amount of application, soil conditions, crop choices. The difficulties in monitoring at farm level and the cost factor make such efforts problematical in most developing countries. The discussion on how far wastewater treatment could be optimized to maintain the desirable nutrient level for downstream irrigation is recent and very little information is available (Dominguez et al., 2004; Drechsel et al., 2004; Jimenez, 2005 ; Martin and Redwood, 2005).

Application of wastewater to either natural or agricultural ecosystems often increases plant biomass production (Jimenez, 1995). However, this increased biomass production and possible deleterious effects on the plant communities within natural ecosystems including changes in species composition and injury and vegetation decline if wastewater is applied in excess. Other adverse effects of wastewater application on vegetation may result from over fertilization, increased disease incidence and depression of biological nitrogen fixation in legumes. Despite these potential problems, reuse of wastewater is a widespread phenomenon worldwide (Jiménez, 2010) and offers substantial advantages over discharge into aquatic ecosystems, especially in arid and semiarid areas of the world.

Vesicular-arbuscular Mycorrhizae in Wastewater Irrigated Soils

An area of concern to us that has been largely overlooked is the potential impact of wastewater nutrients, toxins, pathogens and antimicrobial agents, on the desirable microbiological processes in the soil ecosystems. Of particular interest is the impact on vesicular-arbuscular mycorrhizae (VAM). Since VAM formation and function are sensitive to fertility, the effect of wastewater irrigation on naturally occuring vesicular-arbuscular mycorrhizae is of interest. We know much less about the impact of long-term irrigation with sewage effluent and the soil's buffering properties regulating this impact, especially in even mildly alkaline conditions.

In spite of the important role that vesicular-arbuscular mycorrhizae play in the agricultural ecosystems, seems to be no information on the effect of treated urban wastewater on vesicular-arbuscular mycorrhizas diversity. Only, Safir et al., (1990) and Ortega-Larrocea et al., (2001, 2007) determined the abundance of vesicular-arbuscular mycorrhizae morphotypes in soil irrigated with untreated urban wastewater. However the results cannot be comparable since the untreated wastewater can incorporate significant amounts of pollutants, such as heavy metals in the soil that can depress the vesicular-arbuscular mycorrhizae colonization. Wastewater irrigation by 3 years to an oldfield ecosystem resulted in in a change in spore numbers for the predominate species. Irrigation favored *G. mosseae* over *G. fasciculatum* and increased plant biomass in this study (Safir et al., 1990). Because formation and function of vesicular-arbuscular mycorrhizae are greatly affected by soil nutrient status, especially P and N availability, and because considerable amounts of these nutrients were applied in the wastewater, between 150-270 kg $ha^{-1}yr^{-1}$of N and 30-70 kg $ha^{-1}yr^{-1}$ of P , it was expected that wastewater irrigation would reduce vesicular-arbuscular mycorrhizae in the oldfield ecosystem. Instead, it favored root colonization and increased vesicular-arbuscular mycorrhizae spore population densities in the top soil layer and also deeper in the soil profile (Safir et al., 1990). Ortega-Larrocea et al., (2001) investigate the effect of the accumulated phosphorus and heavy metals introduced into the soils of the Mezquital Valley by >90 years of irrigation have had on the indigenous population of vesicular-arbuscular mycorrhizae. The two types of soil, Leptosol and Vertisol, clearly harbour different sizes of spore population. The abundance of free spores decreased in both soils as a result of 90 years of irrigation with wastewater, and this effect was most marked in the Vertisol. The counts were similar for the two soil types and mainly of the species *Glomus mosseae* and *G. intraradices.* Other species, including *G. geosporum* and few *Acaulospora spp.* were also present. Despite wastewtaer irrigation by 90 years, the concentrations of most of the pollutants do not yet exceed those set down in policy regulations. Irrigation appears to have substantially increased the amounts of P and metals in the soil. The heavy metals in the soil are strongly associated with time under irrigation; the association ranked from strongest to weakest in the following sequence: Cu>Ni>Cr>Zn>P>Pb. Ortega-Larrocea et al. (2007) sampled Leptosols and Vertisols in fields of Mezquital Valley that have been irrigated for four lengths of time 5, 35, 65 and 95 years. In each field they sampled the topsoil around the stalks of maize immediately after harvest. The authors found a steady decline in spore counts with increasing duration of irrigation. Irrigation increased markedly the amount of available P in the soil; it was especially so in the fields irrigated for more than 90 years where the P contents exceeded 3 times the normal crop requirements. The numbers of spores of *G. mosseae* and *G. sinuosa* were at their maxima after only 35 years of irrigation. Root colonization and arbuscule percentages seemed unrelated to the duration of irrigation.

As regards olive processing, olive mill wastewater and olive pomace are the main residual products. In Mediterranean areas such wastewaters are produced over a shortlasting period during the winter time, from November to February , and their amount varies between 7 and 30 millions m^3 per year. The organic fraction of the olive mill wastewater contains a complex consortium of phenolic substances, some nitrogenous compounds (especially amino acids), organic acids, sugars, tannins, pectins, carotenoids, polyphenols and almost all of the water soluble constituents of the olives. The inorganic constituents at the concentration levels found in olive mill wastewtaer are not toxic; quite the reverse, they may potentially serve as good sources of plant nutrients and thereby rendering this effluent potentially suitable for recycling as a soil amendment (Mechri et al., 2011). Olive plants are known to host vesicular-arbuscular mycorrhizae and can be colonized by different species, including *Glomus mosseae*, *Glomus clarum*, *Glomus caledonium*, *Glomus monosporum*, *Glomus intraradices* and *Glomus viscosum* (Calvente et al., 2004). Studies carried out with the olive mill wastewater obtained by the former process of extraction of olive oil have showed that the phenols contained in it were the principal causers of its great antimicrobial activity (Moreno et al., 1987). However, these residues contain several to 10 g of total polyphenols per liter (Mechri et al., 2011), which may suppress plant growth directly and/or through the decrease of AM symbiosis. Depending on concentration, different flavonoids can either increase or decrease AM fungal spore germination, hyphal growth, and hyphal branching.

Actually, although the chemical composition of olive mill wastewater might cause soil and/or water contamination, its use in agriculture as amendment/fertiliser is promoted because of the high content in mineral nutrients, such as N, P, K, Mg and Fe and the consistent rate of organic matter. Di Bene et al., (2013) evaluated the short- and long-term effects of long-lasting repeated olive mill wastewater applications on soil chemical and biochemical properties and arbuscular mycorrhizal fungi. The results showed that olive mill wastewater spreading affected soil chemical properties, microbial biomass carbon, soil respiration, AM fungal root colonisation, arbuscules and vesicles in the short-term period in comparison with control, decreased AM fungal root colonisation.

Pharmaceuticals and personal care products may enter the terrestrial environment through the amendment of agricultural soils with manure or sludge with potential impacts on beneficial soil microbe populations. As such, there is increasing recognition of the need to study the impacts of anthropogenic stressors on plant–microbe interactions in soil ecotoxicology studies and risk assessment. Veterinary and human-use pharmaceuticals in the environment have received increasing attention from the scientific community since the late 1990s, when it was recognized that they were ubiquitous in the environment and that these compounds may be pseudo-persistent due to continuous or repeated input (Halling-Sorensen et al., 1998; Daughton and Ternes, 1999). Research to date has largely focused on the environmental fate, occurrence, and effects of these compounds in aquatic systems. In addition, certain pharmaceuticals have been found to accumulate in soils (Thiele-Bruhn, 2003). Where toxicity testing has been conducted on soil systems, the focus has been on higher organisms such as plants (Migliore et al., 2003; Kumar et al., 2005). Soil fungi have been virtually ignored, despite being critical to ecosystem function as decomposers, mutualists, pathogens and a food-source for many other soil organisms (Hillis et al., 2008). Triclosan (5-chloro-2-[2,4-dichlorophenoxy]phenol; TCS) has been used as a bactericide for over 30 years, and is one of the most common biocides found in pharmaceuticals and personal care products. Triclosan-containing products range from antibacterial mouthwash and toothpaste to household items such as cutting boards, furniture, textiles and sports equipment. Triclosan enters municipal wastewater streams following disposal of TCS-containing consumer products. Twanabasu et al. (2013) using concentrations within the range of those found in North American surface Waters, studied spore germination and hyphal morphology of vesicular arbuscular mycorrhizae exposed to TCS. This study has shown mycotoxic inhibition of the earliest stages of VAM colonization, as well as a reduced response to root signaling compounds. Inhibition of spore germination, hyphal growth, and hyphal branching in non-root wash treatments indicates direct antifungal properties of TCS.

Very little is known about the influence of anthropogenic compounds applied intentionally in the form of pesticides and fertilizers or unintentionally such as pharmaceuticals or personal care products in soil amendments on the structure and function of vesicular arbuscular mycorrhizae. Hillis et al., (2008) evaluate the toxicological potential of a suite of pharmaceuticals regularly found in sludge and manure, on plants, arbuscular mycorrhizal fungi and plant–mycorrhizal interactions using a root-organ culture system. The results from this study suggest that the hyphal length is relatively insensitive until >21 days post exposure. Spore production was even less sensitive, with significant differences typically not appearing until 28 days. A stimulatory effect of the synthetic estrogen, 17-a-ethynyl estradiol, was also observed at small concentrations in the current study. However, at higher concentrations, *D. carota* root length and both *G. intraradices* endpoints were depressed. This response may be due to interference with molecular signaling between plant and fungus and may also explain the effect of carbamazepine on spore production. The readers must read Hillis (2008) to de discution about carbamazepine and 17- a-ethynyl estradiol, tetracycline family, doxycycline. The results of this study suggest that some pharmaceuticals, commonly found in wastewater treatment sludges and livestock manure, exhibit antifungal properties towards vesicular arbuscular mycorrhizae. The pharmaceutical concentration required to cause effects in vesicular arbuscular mycorrhizae were, however, typically much higher than environmental concentrations.

SEWAGE SLUDGE AND ARBUSCULAR MYCORRHIZAS

The sewage sludge as a product of wastewater treatment is considered beneficial because improve soil fertility, biological parameters and plant nutrition. Similar to soil and sediment, the sludge floc is a highly complex microbial system of eukaryotes, bacteria, archaea, and viruses, in which bacteria are dominant and play important roles in removal of organic pollutants and nutrients (nitrogen and phosphorus). The plant benefits particularly through enhanced phosphorus, nitrogen (Cliquet and Stewart, 1993), other mineral nutrients and water uptake, which often results in better growth.

Boyle and Paul (1998) reported mycorrhizal colonization of barley plant either by *G. intraradices* or by *G. mosseae* increased plant dry weight at all sewage sludge application levels.The percentage of root colonization by *G. intraradices* was higher than by *G. mosseae*. The observed decrease in mycorrhizal growth response with increasing sewage sludge is matched with the decline in the percentage of root length colonized as sewage sludge increased. They sugested that the decline in the root colonization with increasing sewage sludge can be attributed to an increase in soil P content. Alguacil et al., (2006) reported that the vesicular-arbuscular mycorrhizae *Glomus intraradices* and *Glomus deserticola* and the addition of composted sewage sludge improve the growth and the N and P contents in shoot tissues of *Juniperus oxycedrus*. Bohner (2009) reported that the percentage vesicular-arbuscular mycorrhizae root colonization varied between different grassland species, even in the same treatment. All grassland species investigated showed a lower percentage of vesicular-arbuscular mycorrhizae colonization of their roots in the sewage-sludge-treated soil than in the unfertilized soil, it seems likely that the much higher inorganic P level in the sewage-sludge-treated soil was associated with a greater P concentration mainly in the roots of the grassland species, leading to a decrease in mycorrhization.

Arriagada et al., (2009) analyzed the effect of the interaction between sewage sludge and *Eucalyptus globulus* nursery seedlings inoculated with arbuscular mycorrhizal fungi as a fertilizers alternative and evaluate its effect on some soil enzyme activities. The seedlings were inoculated with *Glomus claroideum*, *Glomus viscosum*, *Glomus intraradices* and *Glomus constrictum*. The interaction between sludge and *E. Globulus* inoculated with *G. intraradices* increased the vesicular-arbuscular mycorrhizae colonization. Luna et al., (2011) evaluated five comppost prepared with two sludge: from a beverage factory and from a paper industry; they were mixture with organic residues from a pepper caned and corn stuble. Composts were tested over yield of common bean and population of vesicular-arbuscular mycorrhizae. In plots applied with the five composts the number of vesicular-arbuscular mycorrhizae spores found was higher than in control. In all cases spores number were different significatively from those found in the control plot, indicating that there was a positive correlation between soil fertility and the spores number. Seleiman et al., (2013) using sewage sludge application resulted in more numerous mycorrhizal spores in soil and increased root colonization in comparison to synthetic fertilizer. All plants studied had mycorrhizal colonized roots, with the highest colonization rate in maize, followed by hemp. Sewage sludge application resulted in the highest P uptake in all soil-grown plants.

The diversity of glomalean species has been suggested to be a driving factor not only for plant biodiversity but also for ecosystem productivity and stability (van der Heijden et al., 1998), so that loss of fungal biodiversity through human activities may have a negative impact on ecosystem functioning. Sewage sludge spreading, an agricultural practice offering a solution to the problem of waste management, can lead to a reduction in mycorrhizal development in crop plants (Angle and Heckman, 1986; Sainz et al., 1998). However, there is little information about the effects of this human activity on the biodiversity of glomalean populations in agrosystems. Studies using trap cultures of host plants have shown a decrease in number and/or diversity of AM fungal species sporulating in soils amended with sewage sludges (del Val et al., 1999a).

Using Nested PCR based on discriminating primers, Jacquot-Plumey et al. (2001) have been able to monitor AM fungi colonizing *M. truncatula* roots growing in soil from the field plots that received three successive amendments with composted sewage sludges. *G. Claroideum* appeared to be the predominant fungus. Soil amendment with the three composted sewage sludges modified the VAM community. These data underline the diversity between glomalean fungi in their sensitivity towards sewage sludges. Monitoring by PCR of fungal species within roots has revealed that such composted sewage sludges can modify the relative diversity of glomalean fungi making up a natural community, although the extent of their effect will depend on the composition of the sewage sludges and the soil to be amended. The authors concluid, monitoring by PCR of fungal species within roots has revealed that such composted sewage sludges can modify the relative diversity of glomalean fungi making up a natural community, although the extent of their effect will depend on the composition of the sewage sludges and the soil to be amended.

Molecular methods based on 16S rRNA genes and function genes, including fluorescence in situ hybridization, denaturing gradient gel electrophoresis, qRT-PCR, microarray, and proteomics, have been conducted to analyze the microbial community structure and gene diversities at various environments. Jacquot et al., (2000) studied species of arbuscular mycorrhizal fungi in planta and in soil by nested PCR. Nested PCR offers several advantages over these different approaches for monitoring the behaviour of a microcosm fangal community in planta and in soil. The design of taxon-specific primers from the 25S rDNA opens the possibility of discriminating between fungi at the species level directly

in a mixed community. The authors concluid that the negative effect of the sludge on the arbuscular mycorrhizal fungi may be linked to its water-soluble P content whilst that of the enriched sludges, which have the same P content, must result from the higher levels of pollutants in them. Root colonization by *G. mosseae* is enhanced by the presence of the other fungi in the presence or absence of the sludges. Yu and Zhang (2012) studied the microbial community structure and the gene expression levels of an activated sludge from Stanley wastewater tratmen plant in Hong Kong using coupled metatranscriptomic and metagenomic approaches. Bacteria were the dominant domain in both DNA and cDNA datasets, accounting for 92.16% and 68.42% of DNA and cDNA sequences, respectively, while eukaryota comprised approx. 6.43% and 30.97% of total sequences in DNA and cDNA, respectively. Sequences from Archaea and viruses only accounted for 1.10% and 0.18% in DNA sequences, and 0.34% and 0.03% in cDNA, respectively.

Mycorrhizal Interactions with Toxic Metals

Only few plants (the metallophytes) can cope with the adverse conditions on heavy metal soils. Metallophytes have developed various different physiological adaptations which enable them to compete successfully with the harsh conditions in heavy metal soils. Metallophytes are frequently characterized by their ability to hyperaccumulate heavy metals (Regvar et al., 2006). In addition, protection by AMF that colonise plant roots and considerably reduce the uptake of heavy metals into plant cells may be one of the means that allow metallophytes to thrive on heavy metal-polluted sites (Vogel-Mikusˇ et al., 2006). Very recent results obtained from electron-dispersive X-ray spectrometry (EDXA) showed that Zn, Cu and Cd accumulated in the cell wall and in electron-dense granules in the cytoplasm of the fungi while their cytoplasm itself was essentially free of these elements (González-Guerrero et al., 2008). Recently glomalin, an insoluble glycoprotein produced by hyphae of vesicular arbuscular mycorrhizae, was shown to bind potentially toxic elements including heavy metals (González-Chávez et al., 2004). The diversity of AMF spores in heavy metal soils is frequently lower than in non-polluted sites (Regvar et al., 2006). However, a low number of spores does not necessarily reflect a limited AMF development. The description and discuss studies on the expression of genes in plants and fungi under heavy metal stress is detailed by Hildebrandt (2007).

Nazir and Firdaus-Bareen (2011) studied the role of an autochthonous fungal isolate along in association with an vesicular-arbuscular mycorrhizae to enhance the metal uptake in *Helianthus annuus*, a well known high biomass producing plant from metal contaminated soil. Among the five different combinations of soil and tannery sludge, pure sludge was the most toxic. Although the tannery sludge has a reasonable concentration of organic matter but analysis of tannery sludge amended soils showed a high degree of pollution with metals especially Cd, Cr, Na and Zn. In this study, an autochthonous fungus (*Trichoderma pseudokoningii*), obtained from the sludge itself, was used as inoculum alone and in combination with the mycorrhizal fungus *Glomus fasciculatum* and both have shown promising results in aiding the metal uptake in plants. Nakatani et al., (2011) detected the effect of tannery sludge application on AMF spore density, richness, and diversity based on spore population, and on the mycorrhizal status of corn grown in the area. In this study the addition of tannery sludge to agricultural soil decreases vesicular arbuscular mycorrhizae spore density and species richness. However, this practice did not affect mycorrhizal percentage of root colonization rates, indicating that mycorrhizal infective capacity was not impaired in this soil. The authors detected a negative relationship between spore density and some soil nutrients, like phosphorus and mineral nitrogen.

MYCORRHIZOREMEDIATION

The rhizosphere is the soil volume surrounding a root where physical, chemical, and/or biological parameters have been modified by the presence of the root. The rhizosphere is most commonly distinguished from the bulk soil by a depletion of immobile nutrients (NH4+, H2PO4$^-$, and certain micronutrients) and an elevated microbial activity based on readily available carbon derived from root exudation and other root depositions. Rhizosphere technology has been applied to bioremediation of organic pollutants through the use of plants for accelerating degradation rates, and the presence of plants has positive effects on degradation of a wide range of compounds including simple phenolics, aliphatic hydrocarbons, polycyclic aromatic hydrocarbons (PHAs), trichloroethylene, chlorinated aromatics, insecticides, and explosives.

The impact of mycorrhizas, and particularly vesicular arbuscular mycorrhizae, is starting to draw attention in the context of phytoremediation of organics, and recent results have shown that these symbioses may be crucial for both plant establishment on contaminated sites and for degradation of PAHs. A multitude of changes occur in soil in the presence of roots that may be apprehended as changes in its chemical characteristics, modified microbial composition, and enhanced microbial activity (Joner and Leyval, 2003).

The phytoremediation literature is huge and only selected examples related to arbuscular mycorrhizas can be provided here. The reader is referred to several detailed reviews (e.g. Barkay and Wagner-Dobler, 2005). Phytoremediation is an emerging technique that uses living plants for in situ remediation of contaminated soil, sludges, sediments, and ground water through contaminant removal, degradation, or containment. It is an aesthetically pleasing mechanism that can reduce remedial costs, restore habitats, and clean up contamination in situ rather than transporting the problem to another site. In the process of phytoremediation, plants help establish association with mycorrhizal fungi to metabolize the organic pollutants in the rhizosphere. Moreover this mechanism is also supported by secreting plant exudates, e.g. short chain organic acids, phenolics, and small concentrations of high molecular weight compounds (enzymes and proteins) to stimulate bacterial transformations (enzyme induction); by building up of organic carbon to increase microbial mineralization rates (substrate enhancement) or by providing habitat for increased microbial populations and activity. Researchers have examined five plant enzyme systems in sediments and soils. These are dehalogenase enzymes that are important in dechlorination reactions of chlorinated hydrocarbons; nitroreductase which is needed in the first step for degradation of nitroaromatics, laccase which serves to break aromatic ring structures in organic contaminants; whereas peroxidase and nitrilase are important in oxidation reactions. This release of exudates and enzymes that stimulate microbial activity, biochemical transformations and enhancement of mineralization in the rhizosphere has been attributed to mycorrhizal fungi and the microbial consortia. Vesicular arbuscular mycorrhizae induces changes in the host root exudation pattern, which alters the microbial equilibrium in the mycorrhizosphere by influencing host root colonization. It's known to improve plant growth and health by improving mineral nutrition and increasing resistance or tolerance to biotic and abiotic stresses. They also have an ability to alter plant productivity as vesicular arbuscular mycorrhizae can act as biofertilizers, bioprotectants or biodegraders (Khan, 2006). Rhizosphere bioremediation capitalizes on the innate qualities of the vegetation and the potential for mineralization of organic pollutants in the rhizosphere. Therefore, researchers have been focusing on rhizosphere as a zone of enhanced biodegradation and on the plant-microbe interactions that occur there. Rhizosphere bioremediation has been found to increase soil organic carbon, bacteria, and mycorrhizal fungi, all factors that encourage degradation of organics in soil. The mycorrhzospheric microbial community (including non – symbiotic bacteria and fungi) may act in concert with mycorrhizal fungi to degrade organic contaminants. Many persistent organic chemicals like atrazine, hexachlorocyclohexane, aromatic hydrocarbons and polycyclic aromatic hydrocarbons are known to be degraded in the mycorrhizosphere directly or indirectly (Joner & Leyval, 2001; Sainz et al., 2006).

Mycorrhiza not only provide the plants with water and mineral compounds and help to improbé the structure of soil, but have also been shown to act as filters, blocking toxic compounds within their mycelium resulting into reduced toxicity to the plants. Moreover, they influence the physiology of their host plants making them less vulnerable to pathogens, soil pollution, salinity, drought and a number of other environmental stress factors. Mycorrhiza fungal strains have shown improved tolerance to toxic compounds, and hence could increase the success of phytoremediation.

Recently, the enhanced degradation of organic pollutant in soils in the presence of vesicular arbuscular mycorrhizae was observed, indicating that arbuscular mycorrhizal bioremediation is a promising and prospective technique for soils contaminated with organic pollutant. Currently, research is focused mainly on the investigation of mycorrhizal fungi effects on remediation processes. The role of mycorrhizas in phytoremediation was initially attributed to heavy metal plant tolerance and the prevention of erosion. Yet, the widespread occurrence of mycorrhizas, their key role in plant stress tolerance and their impact on other microbial interactions in the rhizosphere, have led to speculation on a possible role in new applications such as degradation of organic pollutants. Małachowska-Jutsz and Kalka (2010) taken soil near the waste lagoons of an existing oil refinery. They collected spores, the morphological identification showed that the dominant genera were Glomus species: *Glomus mosseae*, *Glomus fasciculatum*, *Glomus claroideum*, *Glomus lamellosum.* They concluid that Arbuscular mycorrhizal fungi efficiently contribute to the removal of petroleum contaminants. The diesel contaminant is a type of hazardous fuel materials. Practically, diesel pollutants were difficult to soluble in the water and slow to degrade in the wetland. Tang et al., (2008) studied the effects of an arbuscular mycorrhizal fungus (*Glomus constrictum*) on the growth and some Physio-biochemical indexes of *Zea mays* seedlings under different levels of diesel stress in a pot study. *Glomus constrictum* was isolated from the rhizosphere soil of *Z. mays* in Moxi oil field where the soil had been contaminated by oil. The experimental organic pollutant was diesel #20. This study showed that vesicular arbuscular mycorrhizae could form mycorrhizal symbiont in maize under diesel stress, indicating that mycorrhizal rehabilitation of diesel-contaminated soil was a feasible solution. This study demonstrated that the colonization rate of vesicular arbuscular mycorrhizae on maize root decreased with the increase of diesel concentration in the soil. Phenol and phenolic compounds are the main organic pollutants discharged into petroleum refinery wastewater. Several processes have been used for the elimination of phenolic compounds, with biological treatments being preferred to physicochemical treatments. Several microorganisms which degrade phenolic compounds have been studied, such as the bacterial genera *Acinetobacter*, *Alcaligenes*, *Thauer*a, *Azoarcus*, *Comamonas, Pseudomonas*, *Bacillus* and some yeasts, such as *Candida tropicalis*.

PHOSPHORUS UPTAKE BY PLANTS

Prior to contact, vesicular arbuscular mycorrhizae and their host roots release molecular factors into the rhizosphere. In the case of mycorrhizal interactions, early molecular signals exchanged between the symbionts were identified only recently. Plant roots release strigolactones into the rhizosphere that can stimulate hyphal branching and respiratory metabolism of VAM (Akiyama and Hayashiet al., 2006). These studies suggest an important role for strigolactones in the stimulation of the fungus outside the roots, and possibly also in the progression of VAM hyphae within roots. Reciprocally, VAM fungi release compounds that trigger a variety of responses in plant roots, including calcium spiking, changes in gene expression and lateral root formation (Parniske, 2008). The establishment of the VAM symbiosis can be disturbed by environmental conditions, including P availability which can inhibit the symbiotic interaction (e.g., Breuillin et al., 2010; Ohkama-Ohtsu et al., 2010 Bonneau et al., 2013). Interestingly, recent studies have demonstrated a cross-talk between Pi and nitrogen availabilities to control VAM associations (Bonneau et al., 2013), thus indicating a high level of integration of mycorrhizal responses to mineral nutrition.

Some authors have reported a direct effect of P on VAM spore germination and hyphal growth (de Miranda and Harris, 1994). Evidence has also been gathered for downregulation by P of well established VAM interactions, via a reduced production of root compounds. Balzergue et al. (2011) found that the VAM symbiosis could be arrested almost completely by a high P supply at a very early stage, prior to the attachment of the fungus on the root. The authors also confirmed that the synthesis and exudation of strigolactones are negatively affected by a high availability of Pi, and showed that this effect, like the inhibition of VAM symbiosis, is regulated at the systemic level. Under conditions of P sufficiency, a strong reduction of strigolactone release would prevent the stimulation of VAM fungi and hence the establishment of the interaction. Additional mechanisms targeting the early steps of VAM symbiosis establishment remain to be discovered. Among the mechanisms that could account for the limited root colonization in high P conditions, could include effects on spore germination or hyphal growth (de Miranda and Harris, 1994) .

Balzergue (2013) using germinated spores prestimulated under high P were able to successfully colonize low P-grown roots, indicating that the presence of high P during the spore pre-stimulation stage had not decreased their symbiotic capacity. High P may affect root physiology by altering the hormonal balance. Strigolactones are now recognized as plant hormones involved in several aspects of shoot and root development, and one of the most striking effects of P supply on hormones is a dramatic reduction of root strigolactone synthesis (Balzergue et al., 2011). Strigolactones are known to influence auxin synthesis and transport, thereby modifying root system architecture and root apical meristem function (Koltai and Kapulnik, 2013).

P is an important plant macronutrient, making up about 0.2% of a plant's dry weight. After N, P is the second most frequently limiting macronutrient for plant growth. Pi is also involved in controlling key enzyme reactions and in the regulation of metabolic pathways. The form of P most readily accessed by plants is Pi, the concentration of which rarely exceeds 10 μM in soil solutions. Most studies on the pH dependence of Pi uptake in higher plants have found that uptake rates are highest between pH 5.0 and 6.0, where $H_2PO_4^-$ dominates (Furihata et al., 1992), which suggests that Pi is taken up as the monovalent form. In many agricultural systems in which the application of P to the soil is necessary to ensure plant productivity, the recovery of applied P by crop plants in a growing season is very low, because in the soil more than 80% of the P becomes immobile and unavailable for plant uptake because of adsorption, precipitation, or conversion to the organic form (Holford, 1997).

Therefore, plants must have specialized transporters at the root/soil interface for extraction of Pi from solutions of micromolar concentrations, as well as other mechanisms for transporting Pi across membranes between intracellular compartments, where the concentrations of Pi may be 1000-fold higher than in the external solution. In vascular plants, roots secrete organic acids such as malate and citrate, which compete with Pi for cation binding partners (Johnson and Loeppert 2006) and phosphatases to mineralize Pi from organic compounds. The root epidermis including root hairs increase general surface area and their growth is influenced by Pi availability. Root hair cells participates in direct nutrient uptake from the soil solution at the root-soil interface. Expression of genes encoding high-affinity Pi transporters (PTs) in these cells is maximal in the root apex and root hairs (Gordon-Weeks et al., 2003) and declines in more mature regions. The direct pathway will be most effective immediately behind the apex, where the root epidermis including root hairs, armed with high-affinity PTs, grows into undepleted soil. When plants have an adequate supply of Pi and are absorbing it at rates that exceed demand, a number of processes act to prevent the accumulation of toxic Pi concentrations.

The formation of the arbuscular mycorrhizal symbiosis allows the establishment of a second interface involved in plant nutrient acquisition.

Thus, the arbuscular mycorrhizal symbiosis evolved to improve the nutrient acquisition efficiency of especially Pi, by forming far-reaching extraradical mycelia, which operate as functional extensions of the plant root system (Karandashov and Bucher, 2005). High-affinity phosphate transporters that are expressed exclusively in the extraradical mycelium have been cloned from two AM fungi, *Glomus versiforme* and *Glomus intraradices* (Harrison and van Buuren, 1995). Expression of these transporters is regulated according to the phosphate conditions surrounding the hyphae, and it is likely that they are responsible for phosphate uptake by the extraradical hyphae before transport to the plant. Recognition between symbiotic partners and the establishment of a functional AM symbiosis involve the exchange of many signals, and the molecular basis of early steps in the interaction includes "common symbiosis genes" that are shared with the legume- Rhizobium symbiosis. Examples of VAM-induced genes relevant to P nutrition include the VAM-specific PiTs (Javot et al., 2007). Various physiological situations are known to affect the development of vesicular arbuscular mycorrhizae symbiosis. For instance, plants control the extent to which AM fungi can colonize their roots according to their own nutritional requirements.

Very high P application can certainly alter characteristics of root colonization (particularly reducing arbuscule development) and markedly decrease AM fungal biomass per plant, including both biomass in roots and in soil (Aziz et al., 2011). Bruce et al. (1994) showed that early (up to 15 d) reduction in colonization in cucumber (*Cucumis sativus*) with additional P was mediated by slower growth of fungal infection units within roots, but later there was also a reduction in the rate of formation of new entry points. The latter observation has been significantly extended by Balzergue et al. (2011), who showed marked reductions in appressorium formation on pea (*Pisum sativum*) roots at high P (750 mM; i.e. about 2 orders of magnitude higher than soil solution), which, importantly, was mediated by internal plant-derived signals.

As Pi is the major nutrient delivered by the fungus, the suppressive effect of Pi could be interpreted as a negative regulatory feedback mechanism of the plant host to limit carbohydrate allocation to the symbiont under conditions of optimal Pi supply. This implies that the plant has means to control and limit fungal proliferation according to its nutrient status. The mechanisms involved, however, are unknown. Breuillin et al. (2010) explore the transcriptional changes associated with AM development and with elevated Pi supply in *Petunia hybrida*. Microarray analysis reveals that the symbiosis-associated transcriptome of petunia involves a set of highly conserved genes that overlaps to a large extent with the complement of VAM-associated genes of Medicago, Lotus, and rice. Pi-dependent changes in transcript levels involved mostly the down-regulation of symbiosis-responsive genes encoding PiTs, pathogenesis-related (PR) proteins, and certain proteases. Most interestingly, the suppression of genes encoding enzymes involved in carotenoid and apocarotenoid biosynthesis indicates that these pathways are generally suppressed by Pi. These results suggest that high Pi levels trigger a complex anti-symbiotic syndrome, which results in strong repression of vesicular arbuscular mycorrhizae colonization.

Conclusion

During the last century, a great deal of research and development as well as applications has been devoted to waste. The major reason for such huge efforts is that waste generation constitutes one of the major environmental problems where production industries are concerned. In a new review integrating data from Jiménez and Asano (2008), 46 countries report the use of polluted water for irrigation purposes. When considering wastewater reuse for irrigation an evaluation of the advantages, disadvantages and possible risks has to be made. The reason for the positive as well as for the negative side effects is the high content of organics, nitrogen and phosphorus nutrient as well as the fecal coliforms and debilitating parasites. The high content of organic matter and plant nutrients in the water has improved the physical and chemical conditions of the soils. Soil organic matter increased and so did the crop harvest: the crop yield increased by 94 – 150 %. The irrigated area receives over 80 kg/ha of nitrogen per year. Increasing productivity per unit area, particularly in wastewater contaminated soil, could be achieved by cultivating high yielded cultivar along with importing agronomical practices.

As regards soil microbes, few works focused on arbuscular mycorrhizal fungi. Vesicular arbuscular mycorrhizae form mutualistic symbioses with the roots of most land plants and are fundamental for soil fertility, increasing plant growth, nutrient uptake and improving soil structure and protection against biotic and abiotic stressors (Smith and Read, 2008). Vesicular arbuscular mycorrhizae are widely recognized as components of all terrestrial ecosystems, and known to be essential to above and below-ground processes in ecosystems. Since VAM formation and function are sensitive to soil moisture and fertility, the effect of wastewater irrigation on naturally occuring VAM is of interest.

The responses of mycorrhizal fungi to toxic metals are of importance in view of interest in the reclamation of polluted sites and infiuence on plant growth and productivity. Although there appears to be a wide diversity in responses between different plant-fungal symbioses, vesicular-arbuscular mycorrhizae are usually more tolerant to heavy metals in wastewater than plants without these symbionts (Gaur and Adholeya, 2004; Soares and Siqueira, 2008; Rashid et al., 2009), but mechanisms underlying this protections are largely unknown.

The sewage sludge represents a source of organic matter although its increasing accumulation need a suitable environmental treatment. The sewage sludge as a product of wastewater treatment is considered beneficial because improve soil fertility, biological parameters and plant nutrition. It also improves the soil conditions and plant nutrition through colonization by mycorrhizal fungi as well as reducing leaching and need of synthetic fertilizers. Although there appears to be a wide diversity in responses between different plant-fungal symbioses. Highlighting the need for additional long-term studies on the application of wastewater or sewage sludges in agricultural areas, so that this practice may not impair the biological health of soils. In the specific case of vesicular arbuscular mycorrhizae, it is important to determine doses of sewage sludges that do not reduce the diversity and mycorrhizal inoculum potential, and to assess the symbiotic efficiency of vesicular arbuscular mycorrhizae species exposed to these types of effluents.

Most plants form root symbioses with arbuscular mycorrhizal fungi, which provide them with phosphate and other nutrients. High phosphate availability is known to inhibit the establishment and/or persistence of this symbiosis, but the underlying mechanisms are not understood, there by favoring the direct, non-symbiotic uptake of phosphorus by the root system. The regulation of AM symbiosis by P supply has been observed repeatedly and is considered a general phenomenon. Collectively, the various reports on the down-regulation of AM symbiosis by P suggest that several successive layers of control operate in roots grown under high phosphate. Balzergue et al., (2011) demonstrated for the first time that the regulation of arbuscular mycorrhizal symbiosis by P is accompanied by a systemic regulation of strigolactone production, an important observation with regards to the hormonal function of these compounds. Advances made in the last 10 years have indeed emphasized the importance of signalling events in mycorrhizal interactions, and the recent identification of some signals may shed new light on the regulation of arbuscular mycorrhizal symbiosis by P. Various lines of evidence indicate that AM fungi produce diffusible compounds able to modulate root gene expression (Kosuta et al., 2008), development and metabolism (Gutjahr and Paszkowski 2013). Reciprocally, plant roots secrete compounds that stimulate the fungus. A reasonable hypothesis is that high P availability would decrease the extent of AM symbiosis by reducing strigolactone production in roots (Bouwmeester et al., 2007).

The use of molecular biology technology underlines the interest of using 'omics with taxon-specific primers in order to monitor the effect of abiotic or biotic factors on the different compartments of mycelium development of specific AM fungi introduced into plant production systems. Furthermore, the variability observed in the response of arbuscular mycorrhizal fungi to the sludges makes it possible to envisage their use in ecotoxicological tests to evaluate the quality of wastewtare and sewage sludge for plant production. However, the outcome of such an ecological test may also depend on the host plant used, and this must be borne in mind when proposing a biological system for evaluating wastewater and sewage sludge toxicity.

Finaly, at present, most commercial phosphorus fertilizers on the market such as single superphosphate are derived from phosphate rock. However, world reserves of currently exploitable phosphate rock are rather limited and increasingly dissipated. For these reasons, alternative technologies to recover P from waste materials have received a great deal of interest in recent years. Struvite (MgNH4PO46H2O) is a potentially marketable product for the P fertilizer industry, which can be recovered at municipal wastewater treatment works from anaerobically digested sludge liquors (Gonzalez-Ponce, 2009)

REFERENCES

Ahmed FR, Killham K, Alexander I. (2006) Influences of arbuscular mycorrhizal fungi G. mosseae on growth and nutrition of lentil irrigated with arsenic contaminated water. *Plant Soil* 258:33-41.

Alguacil M, Caravaca F, Diaz-Vivancos P, Hernandez J, Roldan A. (2006) Effect of arbuscular mycorrhizae and induced drought stress on antioxidant enzyme and nitrate reductase activities in *Juniperus oxycedrus* L. grown in a composted sewage sludgeamended semi arid soil. *Plant Soil*, 279:209-218.

Angle, JS, and Heckman, JR (1986) Effect of soil pH and sewage sludge on VA mycorrhizal infection of soybean. *Plant Soil* 93:437-441.

AQUASTAT–FAO Information system on water and agriculture, 1997. http://www.fao.org/ag/aGL/AGLW/aquastat/regions/.

Akiyama K, and Hayashi H. (2006) Strigolactones: Chemical Signals for Fungal Symbionts and Parasitic Weeds in Plant Roots. *Annals of Botany* 97:925-931.

Armienta MA, Morton O, Rodríguez R, Cruz O, Aguayo A, Ceniceros N. (2001) Chromium in a Tannery Wastewater Irrigated Area, León Valley, Mexico. *Bull. Environ. Contam. Toxicol.* 66:189–195.

Arriagada CA, Herrera MA, Garcia-Romera I, Ocampo JA (2004) Tolerance to Cd of soybean (Glycine max) and eucalyptus (Eucalyptus globulus) inoculated with arbuscular mycorrhizal and saprobe fungi. Symbiosis., 36:285-99.

Balzergue C, Puech-Pages V, Becard G, Soizic F, Rochange S. F. (2011) The regulation of arbuscular mycorrhizal symbiosis by phosphate in pea involves early and systemic signalling events. *Journal of Experimental Botany*, Vol. 62, No. 3:1049–1060.

Barkay T. and Wagner-Dobler I. (2005) Microbial Transformations of Mercury: Potentials, Challenges, and Achievements in Controlling Mercury Toxicity in the Environment. *Advances in applied microbiology* Vol. 57:1-52.

Boyle M. and Paul EA (1998) Vesicular-arbuscular mycorrhizal associations with barley on sewage-amended plots. *Soil Biol. Biochem.* 29 No. 6:945-948.

Breuillin F, Schramm J, Hajirezaei M, Ahkami A, Favre P, Druege U, Hause B, Bucher M, Kretzschmar T, Bossolini E, Kuhlemeier C, Martinoia E, Franken P, Scholz U, Reinhardt D (2010) Phosphate systemically inhibits development of arbuscular mycorrhiza in *Petunia hybrida* and represses genes involved in mycorrhizal functioning. *The Plant Journal* 64:1002 1017.

Bohner A. (2009) Effects of sewage sludge application on the arbuscular mycorrhizal colonization of selected grassland species in Austria. International Symposium "Root Research and Applications" RootRAP, 2–4 September 2009, Boku – Vienna, Austria.

Bonneau L, Huguet S, Daniel Wipf D, Pauly N, Hoai-Nam Truong H. (2013) Combined phosphate and nitrogen limitation generates a nutrient stress transcriptome favorable for arbuscular mycorrhizal symbiosis in Medicago truncatula. *New Phytologist* 199:188–202.

Bouwmeester HJ, Roux C, López-Ráez JA, Bécard G. (2007) Rhizosphere communication of plants, parasitic plants and AM fungi. *Trends in Plant Science* 12 (5): 224-230.

Breuillin F, Schramm J, Hajirezaei M, Ahkami A, Favre P, Druege U, Hause B, Bucher M, Kretzschmar T, Bossolini E, Kuhlemeier C, Martinoia E, Franken P, Scholz U, Reinhardt D. (2010) Phosphate systemically inhibits development of arbuscular mycorrhiza in *Petunia hybrida* and represses genes envolved in mycorrhizal functioning. *Plant Journal* 64:1002–1017.

Bruce A, Smith SE, Tester M. (1994) The development of mycorrhizal infection in cucumber: effects of P supply on root growth, formation of entry points and growth of infection units. *New Phytol.* 127:507-514.

Calvente R, Cano C, Ferrol N, Azcón-Aguilar C, Barea JM (2004) Analysing natural diversity of arbuscular mycorrhizal fungi in olive tree (*Olea europaea* L.) plantations and assessment of the effectiveness of native fungal isolates as inoculants for commercial cultivars of olive plantlets. *Appl. Soil Ecol.* 26:11–19.

Cliquet J, Ceorge R, Stewart CR (1993) Ammonia Assimilation in Zea mays lnfected with a Vesicular-Arbuscular Mycorrhizal Fungus *Glomus fasci atum. Plant Physiol.* 101:865-871.

Dalkmann P, Broszat M, Siebe C, Willaschek E, Sakinc T. (2012) Accumulation of Pharmaceuticals, Enterococcus, and Resistance Genes in Soils Irrigated with Wastewater for Zero to 100 Years in Central Mexico. *PloS ONE* 7(9): e45397. doi:10.1371/journal.pone.0045397

Daughton CG, Ternes TA (1999) Pharmaceuticals and personal careproducts in the environment: agents of subtle change? Environ. *Health Perspect.* 107:907–938.

Di Bene C, Pellegrino E, Debolini M, Silvestri N, Bonari E. (2013) Short- and long-term effects of olive mill wastewater land spreading on soil Chemicals and biological properties. *Soil Biology & Biochemistry* 56:21-30.

Domínguez-Mariani E, Carrillo-Chávez A, Ortega A, Orozco-Esquivel M. T. (2004) Wastewater reuse in Valsequillo agricultural area, México. Environmental impact on grounwater. *Water, Air, and Soil Pollution* 155:251–267.

Drechsel P, Giordano M, Gyiele LA (2004) Valuing nutrients in soil and water: concepts and techniques with examples from IWMI studies in the developing world. Research Report 82. International Water Management Institute (IWMI), Colombo, Sri Lanka.

Ensink JHJ, Mahmood T, van der Hoek W, Raschid-Sally L, Amerasinghe FP (2004) A nation-wide assessment of wastewater use in Pakistan: an obscure activity or a vitally important one? *Water Policy* 6:1–10.

Furihata T, Suzuki M, Sakurai H. (992) Kinetic characterization of two phosphate uptake systems with different affinities in suspension-cultured *Catharanthus roseus* protoplasts. *Plant Cell Physiol.* 33:1151-1157.

Gatica J. and Cytryn E. (2013) Impact of treated wastewater irrigation on antibiotic resistance in the soil microbiome. *Environ. Sci. Pollut. Res.* 20:3529–3538.

Gaur A. and Adholeya A. (2004) Prospects of arbusular mycorrhizal fungi in phytoremediation of heavy metal contaminated soils. *Curr. Sci. India* 86:538-534.

González-Chávez MC, Carrillo-González R, Wright SF, Nichols KA (2004) The role of glomalin, a protein produced by arbuscular mycorrhizal fungi, in sequestering potentially toxic elements. *Environ. Pollut.* 130:317–323.

González-Guerrero M, Melville LH, Ferrol, N, Azcón-Aguilar C, Peterson RL (2008) Ultrastructural localization of heavy metals in the extraradical mycelium and spores of the arbuscular mycorrhizal fungus Glomus intraradices. *Canadian Journal of Microbiology.* 54(2):103-110.

Gonzalez-Ponce RG, Lopez-de-Sa E, Plaza C. (2009) Lettuce Response to Phosphorus Fertilization with Struvite Recovered from Municipal Wastewater. *Hortscience* 44(2):426–430.

Gordon-Weeks R, Tong Y, Emyr TG, Leggewis G. (2003) Restricted spatial expression of a high-affinity phosphate transporter in potato roots. *Journal of Cell Science* 116:3135-3144.

Gutjahr C. and Paszkowski U. (2013) Multiple control levels of root system remodeling in arbuscular mycorrhizal simbiosis. *Frontiers in Plant Science Plant Microbe Interaction* 4:1-8.

Halling-Sorensen, B, Nielsen SN, Lanzky PF, Ingerslev F, Holten LüHC, Jorgensen SE (1998) Occurrence, fate and effects of pharmaceutical substances in the environment – a review. *Chemosphere* 36:357–393.

Harrison MJ and Buuren ML (1995) A phosphate transporter from mycorrhyzal fungus *Glomus vermiforme*. *Nature* 378:626-629.

van der Heijden MGA, Klinomonos JN, Ursic M, Moutoglis P, Streitwolf ER, Boller A. (1998) Mycorrhizal fungal diversity determines plant biodiversity, ecosystem variability and productivity. *Nature* 396: 69-72.

Hildebrandt U, Regvar M, Bothe H. (2007) Arbuscular mycorrhiza and heavy metal tolerante. *Phytochemistry* 68:139–146.

Hillis DG, Antunes P, Sibley PK, Klironomos JN, Solomon KR (2008) Structural responses of *Daucus carota* root-organ cultures and the arbuscular mycorrhizal fungus, *Glomus intraradices*, to 12 pharmaceuticals. *Chemosphere* 73:344–352.

Holford ICR (1997) Soil phosphorus: its measurement, and its uptake by plants. *Aust. J. Soil Res*. 35:227–239.

Jacquot-Plumey E, van Tuinen D, Chatagnier O, Gianinazzi S, Gianinazzi Pearson V. (2001) 25S rDNA-based molecular monitoring of glomalean fungi in sewage sludge-treated field plots. *Environmental Microbiology* 3(8):525-531.

Jacquot E, van Tuinen D, Gianinazzi S, Gianinazzi-Pearson V (2000) Monitoring species of arbuscular mycorrhizal fungi in planta and in soil by nested PCR: application to the study of the impact of sewage sludge. *Plant and Soil* 226: 79–188.

Javot H, Penmetsa RV, Terzaghi N, Cook DR, Harrison MJ (2007) A Medicago truncatula phosphate transporter indispensable for thearbuscular mycorrhizal symbiosis. *Proc. Natl. Acad. Sci.* 104:1720-1725.

Jiménez B, Drechsel P, Koné D, Bahri A, Liqa Raschid-Sally L, Qadir M. (2010) Wastewater, Sludge and Excreta Use in Developing Countries: An Overview. In Wastewater Irrigation and Health. Edited by Pay Drechsel, Christopher A. Scott, Liqa Raschid-Sally, Mark Redwood and Akiça Bahri. International Water Management Institute and International Development Research Centre, 2010. ISBN 978-1-84407-795-3.

Jimenez B. (2005) Treatment technology and standards for agricultural wastewater reuse: a case study in Mexico. *Irrig. Drainage* 54 (Suppl.1): S22–S33.

Jimenez B. and Asano T. (2004) Acknowledge all approaches: the global Outlook on reuse. *Water* 21:32–37.

Jimenez B. (1995) Wastewater reuse to increase soil productivity. *Water Science and Technology* 32:173–180.

Joner EJ and Leyval C. (2003) Rhizosphere Gradients of Polycyclic Aromatic Hydrocarbon (PAH) Dissipation in Two Industrial Soils and the Impact of Arbuscular Mycorrhiza Environmental Science and Technology 37(11): 2731-2735.

Johnson SE and Loeppert RH (2006) Role of organic acids in phosphate mobilization from Fe oxide. *Soil Sci. Soc. Amer*. 69: 18-24.

Karandashov V and Bucher M. (2005) Symbiotic phosphate transport in arbuscular mycorrhizas. *Trends in Plant Science* 10(1):22-29.

Keraita B, Drechsel P, Huibers F, Raschid-Sally L. (2002) Wastewater use in informal irrigation in urban and periurban areas of Kumasi Ghana. *Urban Agric. Mag*. 8:11–13.

Khan MA, Shaukat SS, Khan MA (2009) Growth, yield and nutrientcontent of sunflower (*Helianthus annus*) using treated wastewater from waste stabilitation ponds. *Pak. J. Bot.*, 41(3):1391-1399.

Khan AG, Chaudhry TM, Khoo CS, Hayes WJ (2000) Role of plants, mycorrhizae and phytochelators in heavy metal contaminated land remediation. *Chemosphere* 41:197-207.

Koltai H and Kapulnik Y. (2013) Unveiling Signaling Events in Root Responses to Strigolactone. *Molecular Plant* 6 (3):589–591.

Kosuta S, Chabaud M, Lougnon G, Gough C, Denarie J, David G, Barker DG, Guillaume Becard G. A. (2008) Diffusible Factor from Arbuscular Mycorrhizal Fungi induces Symbiosis-Specific *MtENOD11* Expression in Roots of *Medicago truncatula. Plant Physiol.* 131: 952-962.

Kretschmer N, Ribbe L, und Gaese H. Wastewater Reuse for Agriculture. Technology Ressource Management & Development – Scientific Contributions for Sustainable Development, Vol. 2. http://www.sswm.info/sites/default/files/reference_attachments/KRETSC 20et%20al.%202006%20Wastewater%20Reuse%20for%20Agriculure.pdf

Kumar K, Gupta SC, Baidoo SK, Chander Y, Rosen CJ (2005). Antibiotic uptake by plants from soil fertilized with animal manure. *J. Environ. Qual.* 34:2082-2085.

Liu Y and Yongkang Lv Y. (2011) Oxidative damage induced in *Vicia faba* by coke plant wastewater. *Toxicology and Industrial Health* 27(9) 787–792.

Luna Zendejas H, Santos H, Solis M, López W, Vera A, González JM (2011) Effects of compost made with sludge and organic residues on bean (*Phaseolus vulgaris*) crop and arbuscular mycorrhizal fungi density. *African Journal of Agriculture Research* 6(6):1580-1585.

Małachowska-Jutsz A. and Kalka J. (2010) Influence of mycorrhizal fungi on remediation of soil contaminated by petroleum hydrocarbons. *Fresenius Environmental Bulletin Volume* 19(12b):3217-3223.

Martin E. and Redwood M. (2005) Wastewater irrigation in developing countries limitations for farmers to adopt appropriate practices. *Irrig. Drainage* 54 (Suppl. 1):S-63–S-70.

Mechri B, Cheheb H, Boussadia O, Attia F, Mariem FB, Braham M, Hammami M. (2011) Effects of agronomic application of olive mill wastewater in a field of olive trees on carbohydrate profiles, chlorophyll a fluorescence and mineral nutrient content *Environmental and Experimental Botany* 71:184–191.

Migliore L, Cozzolino S, Fiori M. (2003) Phytotoxicity to and uptake of enrofloxacin in crop plants. *Chemosphere* 52:1233–1244.

de Miranda JCC and Harris PJ (994) Effects of soil phosphorus on spore germination and hyphal growth of arbuscular mycorrhizal fungi of arbuscular mycorrhizal fungi. *New Phytol.* 128:103-108.

Moreno E, Perez J, Ramos-Cormenzana A, Martinez J. (1987) Antimicrobial effect of waste water from olive oil extraction plants selecting soil bacteria after incubation with diluted waste. *Microbios* 51:169–174.

Nakatani AS, Mescolotti DLC, Nogueira MA, Martines AM, Miyauchi MYH, Sturmer SL, Cardoso EJBN (2011) Dosage-dependent shift in the spore community of arbuscular mycorrhizal fungi following application of tannery sludge. *Mycorrhiza* 21:515–522.

Nazir A and Firdaus-e-Bereen (2011) Synergistic effect of Glomus fasciculatumand Tricoderma pseudokoningii on Heliantus annus to descontaminate tannery sludge from toxic metals. *African Journal of Biotechnology* 10(22):4612-4618.

Ohkama-Ohtsu N. and Wasaki J. (2010) Recent Progress in Plant Nutrition Research: Cross-Talk Between Nutrients, Plant Physiology and Soil Microorganisms. *Plant Cell Physiol.* 51(8):1255–1264.

Ortega-Larrocea MP, Xoconostle-Caázares B, Maldonado-Mendoza IE, Carrillo González R, Hernández-Hernández J, Díaz Garduño M, López Meyer M, Gómez-Flores L, González-Chávez MC (2010) Plant and fangal biodiversity from metal mine wastes under remediation at Zimapan, Hidalgo, Mexico. *Environmental Pollution* 158:1922–1931.

Ortega-Larrocea MP, Siebe C, Estrada A, Webster R. (2007) Mycorrhizal inoculum potential of arbuscular mycorrhizal fungi in soils irrigated with wastewater for various lengths of time, as affected by heavy metals and available P. *Applied soil ecology* 37:129–138.

Ortega-Larrocea MP, Siebe C, Bécard G, Méndez I, Webster R. (2001) Impact of a century of wastewater irrigation on the abundance of arbuscular mycorrhizal spores in the soil of the Mezquital Valley of Mexico. *Applied Soil Ecology* 16:149–157.

Parniske M. (2008) Arbuscular mycorrhiza: the mother of plant root endosymbioses. Nature reviews Microbiology 6:763-775.

Peasey A, Blumenthal U, Mara D, Ruiz-Palacios G. (2000) A review of policy and standards for wastewater reuse in agriculture: a Latin American Perspective. *WELL Study*, Task No. 68 Part 11

Rashid A, Ayub N, Ahmed T, Gul J, Khan AG (2009) Phytoaccumulation prospects of cadmium and zinc by mycorrhizal plant species grown in industrially polluted soils. *Environ. Geochem. Health* 31:91-98.

Regvar M, Vogel-Mikus K, Kugonic N, Turk B, Batic F. (2006) Vegetational and mycorrhizal successions at a metal polluted site: Indications for the direction of phytostabilisation? *Environ. Pollut.* 144:976–984.

Safir GR, Siqueira JO, Burton TM (1990) Vesicular-arbuscular mycorrhizae in a wastewater-irrigated oldfield ecosystem in Michigan. *Plant and Soil* 121:187-196.

Sainz MJ, Taboada-Castro MT, Vilarino A. (1998) Growth, mineral nutrition and mycorrhizal colonization of red clover and cucumber plants grown in a soil amended with composted urban wastes. *Plant Soil* 205:85-92.

Seleiman MF, Santanen A, Kleemola J, Stoddard FL, Makela PS (2013) Improved sustainability of feedstock production with sludge and interacting mycorrhiza. *Chemosphere.* 91 (9):1236-1242.

Siebe C. (1998) Nutrient inputs to soils and their uptake by alfalfa through longterm irrigation with untreated sewage effluent in Mexico. *Soil Use and Management* 14:119–122.

Siemens J, Huschek G, Siebe C, Kaupenjohann M. (2008) Concentrations and mobility of human pharmaceuticals in the world's largest wastewater irrigation system, Mexico City–Mezquital Valley. *Water research* 2:124-2134.

Smith SE and Read DJ (2008) Mycorrhizal symbiosis. 3rd edn. Academic Press.

Soares CRFS and Siqueira JQ (2008) Mycorrhiza and phosphate protection of tropical grass species against heavy metal toxicity in multi-contaminated soil. *Biol. Fert. Soils* 44:833-841.

Sou-Dakouré MY, Mermoud A, Yacouba H, Boivin P. (2013) Impacts of irrigation with industrial treated wastewater on soil properties. *Geoderma* 200-201: 31–39.

Tang M, Chen H, Huang JC, Tian ZQ (2009) AM fungi effects on the growth and physiology of Zea mays seedlings under diesel stress. *Soil Biol. Biochem.* 41:936-940.

Thiele-Bruhn S. (2003) Pharmaceutical antibiotic compounds in soils – a review. *J. Plant Nutr. Soil Sci.* 166:145–167.

Twanabasu BR, Stevens KJ, Venables BJ (2013) The effects of triclosan on spore germination and hyphal growth of the arbuscular mycorrhizal fungus Glomus intraradices. *Science of the Total Environment* 454–455:51–60.

United Nations New York. 2003. Waste-Water treatment Technologies: A general review. pp 1-122.

del Val C, Barea JM, Azcon-Aguilar C. (1999a) Diversity of arbuscular mycorrhizal fungus populations in heavy-metalcontaminated soils. *Appl. Environ. Microbiol.* 65:718-723.

Vogel-Mikus, K, Pongrac P, Kump P, Necemer M, Regvar M. (2006) Colonisation of a Zn, Cd and Pb hyperaccumulator *Thlaspi praecox* Wulfen with indigenous arbuscular mycorrhizal fungal mixture induces changes in heavy metal and nutrient uptake. *Environ. Pollut.*139;362–371.

Yu K. and Zhang T. (2012) Metagenomic and Metatranscriptomic Analysis of Microbial Community Structure and Gene Expression of Activated Sludge. *PLoS ONE* 7(5): e38183. doi:10.1371/journal.pone.0038183

In: Mycorrhizas
Editor: Elaine Warwick
ISBN: 978-1-63321-436-1

Chapter 7

POTENTIAL OF ARBUSCULAR MYCORRHIZAL FUNGI AND NON-PATHOGENIC FUSARIUM FOR TOLERANCE TO FUSARIUM ROOT ROT AND CHANGES IN ANTIOXIDATIVE ABILITY IN ASPARAGUS DECLINE

***J. Liu*[1], *A. S. M. Nahiyan*[1,2] *and Y. Matsubara*[3*]**
[1]The United Graduate School of Agricultural Science, Gifu University, Gifu, Japan
[2]Advanced Seed Research and Biotech Center, ACI Limited, Dhaka, Bangladesh
[3]Faculty of Applied Biological Sciences, Gifu University, Gifu, Japan

ABSTRACT

Fusarium root rot tolerance and changes in antioxidative ability in arbuscular mycorrhizal fungi (AMF) [*Glomus intraradices* (Gi) and *Gigaspora margarita* (GM)] and non-pathogenic Fusarium (NPF) [*Fusarium oxysporum* f. sp. asparagi (Foa)]-inoculated asparagus plants (*Asparagus officinalis* L., cv. 'Welcome') were investigated. Dry weight

* Corresponding author: ymatsu@gifu-u.ac.jp.

of shoots and roots increased in AMF and NPF-inoculated plants. Though, dry weight increased in NPF-inoculated plants compared than control, which is comparatively lower than AMF-inoculated plants. On the other hand, SOD activity, APX, DPPH radical scavenging activity, content of polyphenol and ascorbic acid increased under inoculation of pathogen stress condition; these effects differed between shoots and roots. As for disease tolerance, the incidence and severity of symptom was less in both AMF and NPF-inoculated asparagus plants than non-inoculated plants, though the effect differed between Gi and GM inoculated asparagus plants with or without presence of NPF. These results suggest that tolerance to Fusarium root rot was increased by AMF colonization with or without NPF in asparagus plants. As well as, antioxidative ability increased in pathogen-stressed conditions, so that the Fusarium root rot tolerance might be associated with the increase in such antioxidative factors.

Keywords: Asparagus decline; Growth promotion; SOD; APX; DPPH

1. INTRODUCTION

Asparagus decline is one of the major constraints in asparagus producing region in the world. This problem is caused by the contribution of both abiotic and biotic factors. Abiotic factors are mainly related to the release of allelopathic compounds [1-4], but also include nutrient imbalance, deterioration of soil physiochemical conditions, cultural factors and excessive harvesting pressure [5-7]. Biotic factors are principally via infestation by soil-borne pathogens that cause crown and root rot, mainly *Fusarium oxysporum* f. sp. *asparagi* (Foa), *Fusarium proliferatum* (Fp) [8-10] and *Fusarium redolens* (Fr) [11].

Asparagus decline is very difficult to control because no resistant cultivar or effective control measure have been developed. Recently, biological control was attempted by inoculation of beneficial microbes, such as arbuscular mycorrhizal fungi (AMF) that had a high affinity for asparagus, and symbiotic relationship promotes vigor in asparagus plants [12]. In addition, Jeffries et al. [13] reported that AMF also act as bioprotectants against pathogen and toxic stresses. Conversely, this effect is not sufficient because many beneficial microbes, such as AMF, *Trichoderma* sp. and *Gliocladium* sp. are inhibited by the toxins, which hinder their ability to promote root health when introduced in the soil [4,14]. Alternatively, certain antagonistic microorganisms can control Fusarium disease in asparagus plants, such as non-pathogenic

Fusarium oxysporum isolates [15]. Blok et al. [16] reported that non-pathogenic *Fusarium oxysporum* (NPF) isolates reduced Fusarium root rot in asparagus plants, as well as several reports have documented the induction of resistance to Fusarium wilt by NPF, as in the case of tomato [17]. However, it was unclear whether the inoculation of NPF enhanced the tolerance to root rot in mycorrhizal asparagus plants.

On the other hand, in pathogen stress condition, production of higher concentration of reactive oxygen species (ROS) such as H_2O_2, superoxide anion (O_2-) and hydroxyl radical (OH) have been shown to create cytotoxic conditions [18]. To overcome this negative consequence of ROS, plants have evolved various protective mechanisms either to reduce or completely eliminate through antioxidative abilities of producing antioxidative enzymes and substances under environmental stresses, such as plant disease, drought and temperature [18,19]. As for mycorrhizal plants, Garmendia et al. [20] reported that disease tolerance and an increase in superoxide dismutase (SOD) activity occurred in pepper, and drought tolerance and antixodative enzymes (activity of SOD and ascrobate peroxidase, APX) increased in mycorrhizal citrus plants [21]. In addition, Zhu et al. [22] demonstrated that tolerance to high temperature stress and an increase in antioxidative enzymes occurred in mycorrhizal maize plants. However, it has been unclear how antioxidative factors changes through AMF and NPF in asparagus plants and how the changes are associated with disease tolerance.

In this study, tolerance to Fusarium root rot and the changes in antioxidative ability were investigated by the inoculation with AMF and NPF.

2. METHODS

2.1. Inoculation of AMF and NPF

Asparagus seedlings (*Asparagus officinalis* L., cv. Welcome) were raised in commercial soil. During the time of sowing, *Glomus intraradices* and *Gigaspora margarita* were inoculated, using commercial inocula (2g/plant) supplied by Idemitsukosan Co., Ltd. and Central grass Co. Ltd. in Japan, respectively; spore densities were unknown. Non-pathogenic *Fusarium oxysporum* (NPF) [*Fusarium oxysporum* f. sp. *asparagi*: NF1] was inoculated (50ml/plant, 10^6 spore/ml) eight weeks after AMF inoculation.

2.2. Inoculation of Foa

Two strains of Foa (MAFF305556 and N9-31) were grown on potato-dextrose agar media. The conidia were harvested in potato sucrose liquid media and incubated at 25°C in the dark for 7 days. The conidial suspension was sieved and its concentration adjusted to 10^6 conidia/ml. Each plant was then inoculated by pouring 50 ml of the conidial suspension on the soil 10 and 14 weeks after AMF inoculation.

2.3. Evaluation of AMF Colonization Level

Eighteen weeks after AM fungi inoculation, roots of asparagus were preserved with 70% ethanol and stained according to Phillips and Hayman [23] to observe the rate of AM fungi-colonized segments in lateral roots (RFCSL). RFCSL express the percentage of 1-cm AM fungi-colonized segments to the total 1-cm segemts of all the lateral roots branched from storage roots of a plant; the number of total segments was approx. 50 per a plant. The average was calculated from the values of four plants.

2.4. Estimation of Symptoms of Fusarium Root Rot

Eighteen weeks after AMF inoculation, the symptoms of Fusarium root rot were rated to 6 degrees as follows: 0, no symptom; frequency of diseased storage roots in a root system: 1, less than 20%; 2, 20-40%; 3, 40-60%; 4, 60-80%; 5, 80-100%. In addition, the disease index was calculated by the following formula:

$$\text{Disease index} = \frac{\Sigma\,(\text{number of plants} \times \text{severity of symptom})}{\text{number of plants} \times 5} \times 100$$

2.5. Analysis of Antioxidative Ability

Eighteen weeks after AMF inoculation, plants were sampled and partitioned into shoots and roots and freezed in liquid nitrogen. Analysis of antioxidative enzyme activities and antioxidative substances were carried out according to the methods of Beauchamp and Fridovich [24] (superoxide dismutase; SOD), Wu et al. [21] (Ascorbate peroxidase), Brurits and Bucar [25] (DPPH radical scavenging activity), Folin and Denis [26] (polyphenol), Roe et al. [27] (ascorbic acid), respectively.

3. Results and Discussion

Eighteen weeks after AMF and 10 weeks after NPF inoculation, dry weight of shoots and roots were increased in AMF-incoculated plants (Figure 1). Dry weight of shoots and roots also increased in NPF-inoculated plants compared than control, which is comparatively lower than AMF-inoculated plants. Hence, plant growth enhancement appeared in mycorrhizal asparagus plants. AMF colonization occurred in all the inoculated plots, and RFCSL showed highest value comparatively Gi inoculated plot among the treatments (Figure 2).

As for disease tolerance, reddish brown root lesions appeared in all the treatments 18 weeks after pathogenic Foa inoculation, and the severity of symptom was less in both mycorrhizal and NPF-inoculated asparagus plants than non-inoculated plants (Figure 3). In that case, we also observed severity of symptom was also lower in the combination of AMF and NPF treatments, same phenomena was also observed in disease indices, though it varies between two pathogenic Foa (MAFF305556 and N9-31). However, the disease indices revealed that the severity was remarkably low in Gi-inoculated asparagus seedling with or without presence of NPF.

In this study, growth promotion was observed by the inoculation of AMF and NPF. Matsubara et al. [28] reported that dry weight became greater in AMF inoculated asparagus plants, which is very similar to our study. The incidence and severity of root rot symptom in asparagus seedlings were suppressed by both AMF and NPF. Matsubara et al. [29] reported that Fusarium root rot tolerance increased in mycorrhizal asparagus plants as well as, Blok et al. [16] reported that severity of Foa decreased by applying NPF. Same phenomenon was also observed in tomato [15,17].

As for antioxidative ability in this study, activities of antioxidative enzymes (SOD, APX) and DPPH radical scavenging activity increased under inoculation of pathogen among some of the plots (Figure 4). Same phenomenon was also observed in polyphenol content and ascorbic acid content. Similar pattern was observed in strawberry plants after inoculation of *Fusarium oxysporum* f. sp. *fragariae* and *Colletotrichum gloeosporioides* [30].

SOD acts as defensive reactions and detoxify superoxide (O2-) among the antioxidative enzymes; thus, SOD activity is considered the most important key enzyme in antioxidative abilities in plants [31]. Sahoo et al. [18] mentioned that SOD activity increased in *Phytopthora* blight in taro under induced blight condition compared with control. In mycorrhizal pepper plants, increase in SOD activity and disease tolerance appeared in pathogen (*Verticillium dahliae*) stress condition [20]. Moghaddam et al. [19] mentioned that SOD activity was higher in a resistant strawberry cultivar than susceptible cultivars with *Mycosphaerella fragariae* infection. From these findings, antioxidative ability might be closely related with disease tolerance in AMF and NPF plants.

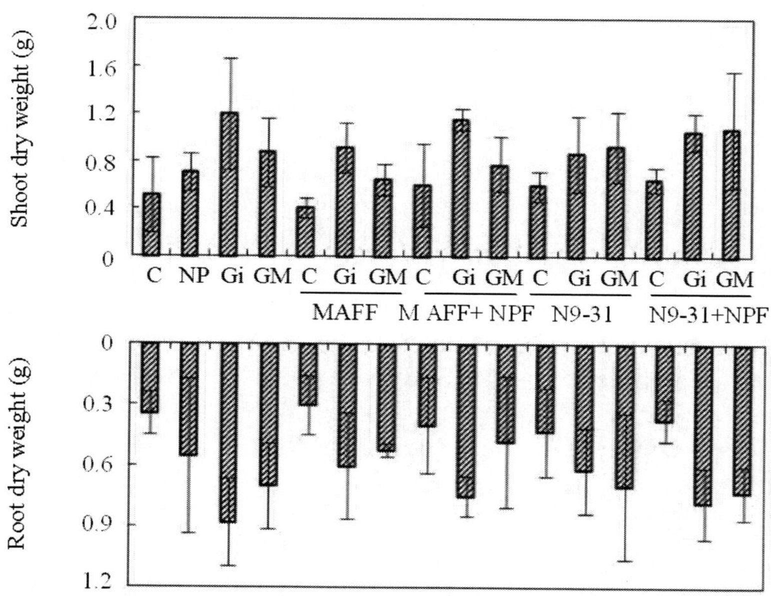

C, AMF-non-inoculated; Gi, *Glomus intraradices*; GM, *Gigaspora margarita*; NPF, non-pathogenic Fusarium; MAFF and N9-31, *Fusarium oxysporum* f. sp. *asparagi*.

Figure 1. Asparagus plant growth 18 weeks after AMF inoculation.

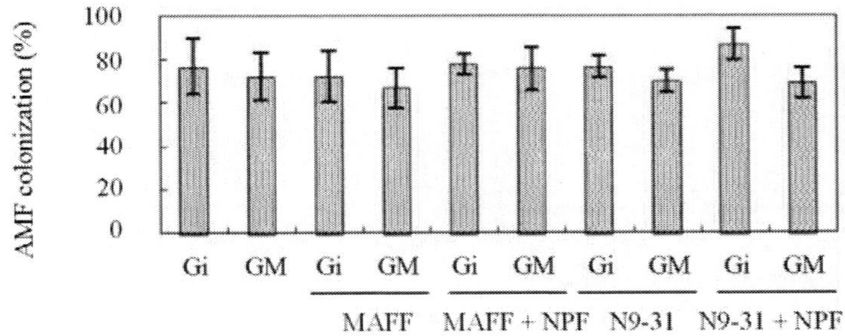

Gi, GM, NPF, MAFF and N9-31, see Figure 1.

Figure 2. AMF colonization in asparagus plants 18 weeks after AMF inoculation.

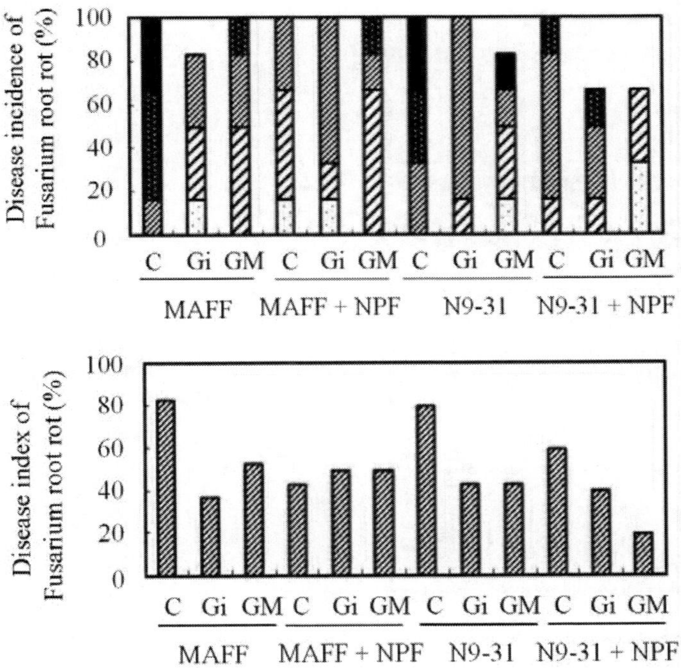

C, Gi, GM, NPF, MAFF and N9-31, see Figure 1. □20%; ▒ 20-40%;, ▨ 40-60%; ▩ 60-80%, ■ 80-100%.

Figure 3. Disease incidence and index of Fusarium root rot in asparagus plants 18 weeks after AMF inoculation.

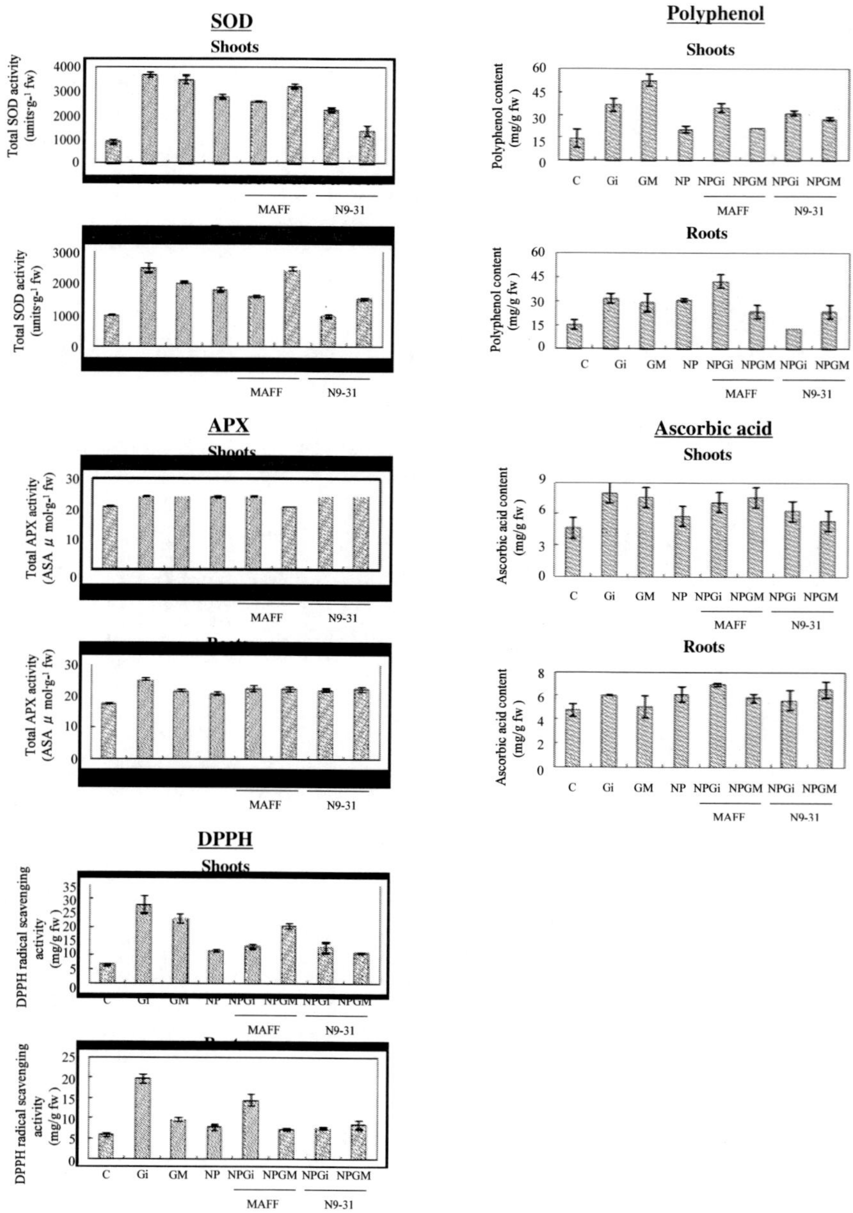

C, Gi, GM, NPF, MAFF and N9-31, see Figure 1.

Figure 4. SOD, APX, DPPH radical scavenging activity, polyphenol and ascorbic acid contents in shoots and roots 18 weeks after AMF inoculation.

From the facts in this study, tolerance to Fusarium root rot was increased by AMF and NPF in asparagus plants. These findings might be helpful to mange the disease and improve plant health in asparagus decline.

Conclusion

In this study, tolerance to Fusarium root rot was increased by AMF colonization with or without NPF in asparagus plants. As well as, antioxidative ability increased in pathogen-stressed conditions, so that Fusarium root rot tolerance might be associated with the increase in such antioxidative factors.

Acknowledgments

This study was supported by Grants-in-Aid for Scientific Research (No. 21580029), Japan Society for the Promotion of Science (JSPS).

References

[1] Yang, H. (1982). Autotoxicity of *Asparagus officinalis* L. *Journal of American Society for Horticultural Science*, *107*, 860-862.

[2] Hartung, A. C. & Stephens, C. T. (1983). Effect of allelopathic substances produced by asparagus on incidence and severity of asparagus decline due to *Fusarium* crown rot. *Journal of Chemical Ecology*, *9*, 1163-1174.

[3] Hazebroek, J. P., Garrison, S. A. & Gianfagna, T. (1989). Allelopathic substances in asparagus roots: extraction, characterization, and biological activity. *Journal of American Society for Horticultural Science*, *114*, 152-158.

[4] Blok, W. J. & Bollen, G. J. (1993). The role of autotoxins from root residues of the previous crop in the replant disease of asparagus. *Netherland Journal of Plant Pathology*, *99* (Supp. 3), 29-40.

[5] Hamel, H., Vujanovic, V., Nakano-Hylander, A., Jeannotte, R. & St-Arnaud, M. (2005). Factors associated with fusarium crown and root rot of asparagus outbreaks in Quebec. *Phytopathology*, *95*, 867-873.

[6] Yergeau, E., Vujanovic, V. & St-Arnaud, M. (2006). Changes in communities of *Fusarium* and arbuscular mycorrhizal as related to different asparagus cultural factors. *Microbial Ecology*, *52*, 104-113.

[7] González, M. I. & Del Pozo, A. (2008). Harvest pressure in green asparagus. Proc. XIth International Symposium on Asparagus. *Acta Horticulturae, 776*, 75-80.

[8] Schreuder, W. & Lamprecht, S. C. (1995). Pathogenicity of three *Fusarium* species associated with asparagus decline in South Africa. *Plant Disease*, *79*, 177-181.

[9] Yergeau, E., Filion, M., Vujanovic. V. & St-Arnaud, M. (2004). A PCR-denaturing gradient gel electrophoresis approach to asses *Fusarium* diversity in asparagus. *Journal of Microbiology Methods*, *60*, 143-154.

[10] Wong, J. Y. & Jeffries, P. (2006). Diversity of pathogenic *Fusarium* populations associated with asparagus roots in decline soils in Spain and the UK. *Plant Pathology*, *55*, 331-342.

[11] Baayen, R. P., Van der Boogert, P. H. J. F., Bonants, P. J. M., Poll, J. T. K., Blok, W. J. & Waalwijk, C. (2000). *Fusarium redolens* f. sp. *asparagi,* causal agent of asparagus root rot, crown rot and spear rot. *European Journal of Plant Pathology*, *106*, 907-912.

[12] Matsubara, Y., Harada, T. & Yakuwa, T. (1994). Effect of vesicular-arbuscular mycorrhizal fungi inoculation on seedling growth in several species of vegetable crops. *Journal of Japanese Society for Horticultural Science*, *63*, 619-628.

[13] Jeffries, P., Gianinazzi, S., Perotto, S., Turnau, K. & Barea, J. (2003). The contribution of arbuscular mycorrhizal fungi in sustainable maintenance of plant health and soil fertility. *Biology and Fertility of Soils*, *37*, 1-16.

[14] Pedersen, C. T., Safir, G. R., Siqueira, J. O. & Parent, S. (1991). Effect of phenolic compounds on asparagus mycorrhizae. *Soil Biology and Biochemistry*, *23*, 491-494.

[15] Da Silva, J. C. & Bettiol, W. (2005). Potential of non-pathogenic *Fusarium* isolates for control of Fusarium wilt of tomato. *Tropical Plant Pathology, 30*, 409-412.

[16] Blok, W. J., Zwankhuizen, M. J. & Bollen, G. J. (1997). Biological control of *Fusarium oxysporum* f. sp. *asparagi* by applying non-pathogenic isolates of *F. oxysporum. Biocontrol Science and Technology*, *7*, 527-541.

[17] Fuchs, F. G., Loccoz, Y. M. & Defago, G. (1999). Ability of nonpathogenic *Fusarium oxysporum* Fo47 to protect tomato against Fusarium wilt. *Biological Control, 14*, 105-110.

[18] Sahoo, M. R., Dasgupta, A. M., Paresh, A., Kole, C., Jayant, A., Bhat, S. & Mukherjee , A. (2007). Antioxidative enzymes and isozymes analysis of taro genotypes and their implications in *Phytophthora* blight disease resistance. *Mycopathologia, 163*, 241-248.

[19] Moghaddam, B., Charles, M. T., Carisse, O. & Khanizadeh, S. (2006). Superoxide dismutase responses of strawberry cultivars to infection by *Mycosphaerrella fragariae*. *Journal of Plant Physiology, 163*, 147-153.

[20] Garmendia, I., Aguirreolea, J. & Goicoechea, N. (2006). Defence-related enzymes in pepper roots during interactions with arbuscular mycorrhizal fungi and/or *Verticillium dahliae*. *Biocontrol, 51*, 293-310.

[21] Wu, Q., Zou, Y. & Xia, R. (2006). Effect of water stress and arbuscular mycorrhizal fungi on reactive oxygen metabolism and antioxidant production by citrus (*Citrus tangerine*) roots. *European Journal of Soil Biology, 42*, 166-172.

[22] Zhu, X., Song, F. & Xu, H. (2010). Influence of arbuscular mycorrhiza on lipid peroxidation and antioxidative enzyme activity of maize plants under temperature stress. *Mycorrhiza*, *20*, 325-332.

[23] Phillips, J. M. & Hayman, D. S. (1970). Improved procedure for clearing roots and staining parasitic and vesicular-arbuscular mycorrhizal fungi for rapid assessment infection. *Transactions of British Mycological Society, 55*, 158-163.

[24] Beauchamp, C. & Fridovich, I. (1971). Superoxide dismutase: Improved assays and an assay applicable to acrylamide gels. *Annals of Biochemistry, 44*, 276-287.

[25] Burits, M. & Bucar, F. (2000). Antioxidative activity of *Nigella sativa* essential oil. *Phytotherapy Research, 14*, 323-328.

[26] Folin, O. & Denis, W. (1915). A colorimetric method for the determination of phenols (and phenol derivatives). in urine. *Journal of Biological Chemistry, 22*, 305-308.

[27] Roe, J. H., Mary, B. M., Oesterking, M. J. & Charlotle, M. D. (1948). The determination of diketo-L-gluconic acid, dehydro-L-ascorbic acid and L-ascorbic acid in the same tissue by 2,4-dinitrophenyl hydrazine method. *Journal of Biological Chemistry, 174*, 201-208.

[28] Matsubara, Y. & Harada, T. (1996). Effect of vesicular-arbuscular mycorrhizal fungi infection on growth and mineral nutrient content. *Journal of Japanese Society for Horticultural Science, 65*, 603-609.

[29] Matsubara, Y., Hasegawa, N. & Ohba, N. (2003). Relation between fiber and pectic substances in root tissue and tolerance to Fusarium root rot in asparagus plants infected with arbuscular mycorrhizal fungus. *Journal of Japanese Society for Horticultural Science*, *72*, 275-280.

[30] Li, Y., Yanagi, A., Miyawaki, Y., Okada, T. & Matsubara, Y. (2010). Disease tolerance and changes in antioxidative abilities in mycorrhizal strawberry plants. *Journal of Japanese Society for Horticultural Science*, *79*, 174-178.

[31] Fridovich, I. (1986). Biological effects of the superoxide radicals. *Archives of Biochemistry Biophys*, *247*, 1-11.

INDEX

B

C

F

G

H

I

J

K

L

M

N

O

P

Q

R

S

T

U

V

W

X

Y

Z